Thomas Hentrich
Wilhelm R. Rossak

Efficient Service Discovery in Large-scale Mobile Agent
Systems

Thomas Hentrich
Wilhelm R. Rossak

Efficient Service Discovery in Large-scale Mobile Agent Systems

A Self-adaptive Infrastructure
for Robust Communication in Highly
Distributed Software Systems

VDM Verlag Dr. Müller

Imprint

Bibliographic information by the German National Library: The German National Library lists this publication at the German National Bibliography; detailed bibliographic information is available on the Internet at http://dnb.d-nb.de.

Any brand names and product names mentioned in this book are subject to trademark, brand or patent protection and are trademarks or registered trademarks of their respective holders. The use of brand names, product names, common names, trade names, product descriptions etc. even without a particular marking in this works is in no way to be construed to mean that such names may be regarded as unrestricted in respect of trademark and brand protection legislation and could thus be used by anyone.

Cover image: www.purestockx.com

Publisher:
VDM Verlag Dr. Müller Aktiengesellschaft & Co. KG, Dudweiler Landstr. 125 a, 66123 Saarbrücken, Germany,
Phone +49 681 9100-698, Fax +49 681 9100-988,
Email: info@vdm-verlag.de

Produced in USA and UK by:
Lightning Source Inc., La Vergne, Tennessee, USA
Lightning Source UK Ltd., Milton Keynes, UK
BookSurge LLC, 5341 Dorchester Road, Suite 16, North Charleston, SC 29418, USA

ISBN: 978-3-639-01385-6

*To my parents
and grandparents*

Acknowledgments

I would like to thank all the people who contributed to this book. The support that I received from mentors, friends, and my family has been invaluable for success of this work.

Special thanks to Prof. Dr. W. R. Rossak, for guiding me through the curriculum since the very beginning of my studies. If there was anything that ultimately arouse my excitement for the field of Computer Science, it was his way of teaching. His support over the past years has been key for several cornerstones of my career.

I further wish to thank Dr.-Ing. Arndt Döhler for many worthwhile discussions. His efforts in this field of research greatly inspired the work described herein, and he successfully demonstrated for the first time the usability of the developed software components under real-life conditions within the Quick-LinkNet framework.

My thanks also go to my former colleagues Erin Hill and Daniel Cooke, as well as to my brother Andreas for spending so much time with me to get the document compiled. And finally, from the bottom of my heart, I wish to thank Julia. For everything. I love you.

Abstract

Mobile agent technology represents one of the most promising concepts to describe and implement large-scale distributed software systems and communication for highly dynamic network environments. Such systems achieve their application functionality through mobile software entities, so-called agents, which migrate between networked computers and use provided services. As a result, mobile agents must identify and navigate to suitable nodes in order to fulfill their application-specific tasks. In particular, they must be able to discover services independently and without any human intervention. Autonomous planning and navigation, in turn, require efficient handling of large amounts of service descriptions and routing information from the underlying network. Existing infrastructures, however, are challenged by this type of software architecture as they do not scale with the volume of information, underlying network dynamics, and frequency of changes.

Against this background, a novel infrastructure supporting mobile agent systems, QuickLinkNet, is being developed at the University of Jena, Germany. QuickLinkNet represents a multi-tier middleware that can cope with the demands of modern service-oriented systems in distributed environments. This book introduces QuickLinkNet's top tier that is represented by the APLI-COOVER software component. This component employs peer-to-peer technology to establish a self-adapting and fault-tolerant software layer that allows building distributed applications at a global scale and manages the arising communication volume in mobile agent systems efficiently.

Contents

1 Introduction

This book addresses mobile agent technology and the challenges that arise with service-oriented migration strategies. In particular, the infrastructural aspects represent the focal point of interest hereinafter.

Mobile agent systems exhibit their capabilities especially well in distributed network environments in which they use data sources and application services to fulfill user-given tasks. Service-oriented migration terms their capability to autonomously discover these sources and navigate towards them. However, besides the agents themselves, the underlying agent system and its infrastructure play a crucial role, too.

Motivated by [Erf04, Doe05], who postulate that mobile agent systems demand for decentralized and self-organizing infrastructures to cope with the continuously growing amount of data sources and application services, this work regards peer-to-peer systems as a promising technology to establish this basis.

The following chapter introduces mobile agent technology and reveals the challenges of service discovery in detail. In chapter 2, characterizing aspects of peer-to-peer technology are highlighted, before chapter 3 depicts three selected implementations in depth. Chapter 4 finally presents APLICOOVER, the contribution of this work towards the QuickLinkNet [1] infrastructure framework for mobile agent systems that is being developed at the University of Jena, Germany.

1.1 Mobile agent technology

Research on software agents has become a vibrant topic over the past years in many disciplines related to information technology. However, each discipline has a different understanding of the notion, and agents appear in various types. They can be found throughout the spectrum, ranging from sophisticated applications in Artificial Intelligence to embodied entities for deep space explorations [Acq02]. Sometimes they are objects of research on their own, sometimes they are rather mediators for other technologies.

The first concepts of software agents already appeared in the mid-1970s in the domain of Artificial Intelligence [HCB77]. Within this research domain, agents are computer programs that appear intelligent in a certain way. Intelligent means that these programs are able to learn and to adapt their behavior. To a certain degree, they also have to be independent and be able to perform some task on behalf of the user. If these characteristics hold, a computer program can be called *agent*.

---

[1] http://swt.informatik.uni-jena.de/Projekte/QuickLinkNet.html

They are mentioned first in the mid-1990s [Bra97], when White introduced *Telescript* as a dedicated programming language and runtime environment for mobile agents. Mobile agents mainly differ from standard software agents by their additional capability to move autonomously within a computer network. That means mobile agents can hop between different devices.

This work concentrates on mobile agents and their application in Software Engineering. Within this context, mobile agents are understood in a rather pragmatic sense. Their intelligent aspect, as it is emphasized in research on Artificial Intelligence, is neglected. For software engineers mobile agents represent a high-level abstraction to describe and effectively implement distributed applications. With mobile agents, completely new applications in the software landscape, which made no sense in earlier evolutionary stages, are now conceivable. From that point of view, mobile agent technology can be understood as a new paradigm in software design. It provides promising particularities and tackles several weak points of traditional software systems [Lin01].

The following sections will show first where exactly mobile agent technology represents a new paradigm. Out of this description one of the concept's major challenges is revealed and this work's contribution towards a solution is outlined.

1.1.1 Looking back in history

To understand the underlying concepts of software systems, the surrounding influences and the historical development have to be considered. Looking back to the early stages of silicon-based computing, single mainframes, scattered workstations, and poorly equipped terminals ruled the scenery. Computer networks first emerged in the late 1960s. What we call the Internet today started in 1969 as a network of four computers in the USA. It was named *ARPANET* after the *Advanced Research Projects Agency*, a subdivision of the *Department of Defense*, that initialized its development.

It took another ten years until the first personal computers entered the stage. The associated decay in hardware prices and the increasing demand of computational power led to an acceleration in development and a huge spread of micro processor systems in the 1980s. Soon, these systems also gained access to the Internet and the various services enjoyed growing popularity. The final breakthrough came with the introduction of the World Wide Web (WWW) service that facilitated searching and navigating in the growing amount of data. The Internet virtually exploded in the mid-nineties as it finally became a world-wide means of communication, exchange, and trade. Rapidly evolving technology and exponential growth of accessible data quickly brought us to a point where hundreds of millions of people use the Internet from everywhere at every time. Today, any kind of information, from e-mails and images to audio data and video steams, is being transferred via the Internet. It has penetrated our everyday lives at all levels, created new markets, and tremendously influences culture.

In summary, the development of computing can be seen as a continuous shift from solitary computers, to connected groups, and finally to a ubiquitous network infrastructure.

In the history of software systems, a similar evolution can be found. In the 1960s, higher-level programming languages had just overcome their infancy and were just about to mature and diversify. Applications were rather monolithic and fairly interacted with users directly. Computational tasks were queued as batch jobs and processed sequentially. Later, the more computers were connected, the more the client-server paradigm found its way into system architecture. This concept assumes a clear separation of roles. One normally finds several clients that use services an associated server provides. Probably the most familiar example representing this idea is the WWW where numerous clients request web pages from the server hosting the files. Usually, the user requests a web site because he or she is looking for a particular piece of information. If, for instance, the user is trying to find the cheapest flight offer for a certain destination, dedicated flight auctions and special search engines are good places to start searching. The user will most likely visit several such web sites sequentially, fill out the required search forms, note the best offers, finally pick one flight, and book it. Fig. 1.1 visualizes the process. A closer

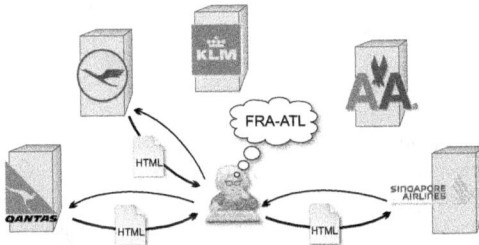

Figure 1.1: Searching for flights, using a client-server application.

look shows first, the user has to know concrete sources of information on where to find flight offers: here these sources are most likely flight portals and search engines. Within the overwhelming, continuously growing, and changing amount of web pages, this is at least a time-consuming burden. In addition, not all services might be accessible via the WWW. Web Services, for example, pursue a different approach of information lookup and data exchange, and find more and more acceptance in business environments [Boo04, Jab03, Küs03].

Second, while browsing through the different sites, every single web page in the above sketched situation has to be transferred from the Internet servers to the client running the web browser. Most of them are useless for the user in the example. Nevertheless, each page has to be transferred entirely because only the user can decide what is actually relevant and what is not. There is no intelligent means in between the server and the browser that assists the user in this selection process. Another obvious drawback of web-based search in the flight booking scenario is that the itinerary information has to be entered recurrently on the different web forms. The sooner the user gets bored by this procedure or the less time he or she wants to spend searching, the more likely it becomes that the best flight has not been found yet.

In brief, the current Internet is mainly an information repository for human consumption [Fei05]. Users are heavily involved in any kind of information lookup, filtering, and processing of data. This might be because the remote data lacks structure and can't be interpreted by a program, more frequently, however, the reason is that there exists no efficient concept to filter the data.
Of course one could utilize filter programs on the client side, gathering remote data and sieving out relevant portions. This is exactly how web crawlers, search bots, and spiders work. One could also use dedicated interfaces on the server that would accept complex queries. But both approaches have strong drawbacks: In the first case, the user is truly disburdened and can switch its attention to other things, since a program would filter the remote data. Form a technical perspective, however, this is an immense waste of network bandwidth and unnecessary server load. In the second case, a dedicated server interface would require a matching caller mechanism on the client side to use that interface appropriately. This specialization results in an even stronger dependency between a client and a server. That might lead to a niche where none of them can switch to alternatives, since the client requires this special type of server, and the server can only serve this special group of clients. In light of the fast-moving IT landscape this too is a questionable approach.
Nevertheless one could at least accept the first approach as long as the client device is powerful enough and its network connection is broad and stable. But nowadays, these conditions can't be simply taken for granted. Notebooks, PDAs, and even cell phones have joined the group of client devices. They no longer rely on permanent connections, but have gained wireless access to the Internet. They often connect only for a short time causing much higher dynamics within the network. Additionally, their connections are less stable and reduced in bandwidth. In many cases the devices have limited hardware resources, like small displays and slow processors, making it difficult to process huge amounts of data and capture relevant information. It is apparent that this new environment challenges traditional approaches and technologies. In some cases it even becomes completely impossible to efficiently utilize them what emphasizes the necessity for new architectural paradigms in this software domain.

1.1.2 Outlining the paradigm

This is where agent technology comes into play—to be more precise, mobile agent technology. Mobile agents are software components, consisting of algorithms and data, which can be exchanged over a computer network [Mül97]. First of all, this description sounds close to some object-oriented ideas. Objects already offer wrapping of algorithms and data, and with object serialization there are even concepts to move software entities within a network [Her03, Pit01]. To reveal the actual novelty of mobile agent technology and to see to what extent agents are superior to regular objects, looking at a higher level of abstraction proves helpful. Further details are examined in section 1.1.3. On the abstract level, the notion of the agent in its original, human-related understanding is emphasized. The Mobile Agent Facility Specification [Bel00], for instance, characterizes an agent as a computer program that acts autonomously and that is able to control its own thread of execution. Along with Wikipedia's definition [Wik06] the concept can be further tightened. Thereafter an agent is:

- *autonomous* – capable of task selection, prioritization, goal-directed behavior, and decision-making without human intervention,

- *proactive* – able to decide on itself when to perform some activity,

- *reactive* – able to perceive the context in which it operates and react to it appropriately; and

- *social skilled* – able to engage other software components through some sort of communication and coordination to collaborate.

Finally, if an agent can even move within a network and thereby select its site of execution, it is called a *mobile agent* [Bra05]. There are several other characterizations, each one of them strengthening particular aspects of the notion. In [Fra96], for instance, adaption, flexibility, and even personality are additionally attributed to agents. To adapt to new situations, to derive some appropriate behavior from past experience, and to develop own identity, those agents have some sort of representation of history and of their surroundings. They use this knowledge to decide how to react in a particular environment. These aspects accentuate the idea of an agent's intelligence. In a simple case, this intelligence are functions monitoring parameters of the environment and triggering some appropriate agent behavior when certain thresholds are reached. Advanced methods use competing behaviors, versatile control circuits, and neural networks. Agents that can adopt, learn, and reason—at least in a figurative meaning—are called *intelligent agents* [Ben96].

The definition so far might tempt one to look at agents in the wrong shade of light, since there are some misleading allegorical parallels with their human counterparts. Although agents are able to perform tasks autonomously, have social skills and personal traits, can jump between different computers, and possess certain intelligence, at the present time they have neither a free will nor any set of moral principles or values. Software agents are computer programs that obediently assist their users in computational work. Any association between software agents and computer viruses or human agents as investigators are completely unsubstantial. Software agents do not rummage unrestrainedly and they do not spy personal data. They 'live' in a secure environment and can be fully controlled. Moreover, it should be emphasized again that all aspects of intelligence and autonomy of mobile agents refer to intelligent and autonomous migration in a network. Intelligent and autonomous behavior in a broader sense, as it is addresses with research on intelligent agents, is explicitly

5

left out. This strengthens the above mentioned pragmatic character of the mobile agent concept and its application in software engineering.

In the sense of this work, a mobile agent is a building block of a software system and has to be seen as an assistant of the agent application's user. The agent is handed over a task and tries to solve the problem instead of the user. The given task thereby serves as a logical boundary. Only within this logical fence the agent's intelligence and its own personal characteristics come to light. They focus on intelligent service discovery and efficient utilization of the discovered services. The vision is that the user only has to state *what* task should be performed, and the agent is able to derive which services in the network must be employed to accomplish that task. That means, the agent is responsible for handling the *how* of the task.

Using mobile agents, the flight booking example from above might be re-sketched as follows: The user is no longer bound to its desktop computer and its web browser. Any device that is capable of hosting agents and that allows entering data can be used to hand over the itinerary information to a booking agent. The user's cell phone for instance should be sufficient. As soon as the task has been assigned to the agent, the user's part is done and the agent starts working. Probably, the booking agent can rely on a list of aviation-related servers from previous search jobs. If not, querying a directory might be a good point to start searching for promising services. In either case, the agent will most likely leave the cell phone, access the Internet at some dedicated entry point, and try to reach the first server of interest. On its way through the network, as it is visualized in Fig. 1.2, the agent migrates between different computers, visits the servers from its list, and gathers their flight offers. At any time, flight advertisements, new discovered services, and orphaned hosts might lead to dynamic adjustments of its route. When querying offers on some server, the agent does not blindly take all of them. Instead, the agent will compare each new offer to the ones already gathered on previously visited servers, and will keep only those that best fit the user's demands. After a certain amount

Figure 1.2: Searching for flights, using a mobile agent application.

of time, or as soon as enough offers have been found, the agent will return to its principal. In the meantime, however, the latter might have switched off its cell phone, forcing the agent either to wait on an intermediate host or to jump to one of the user's different devices. Assuming that there is such a device and the agent is able to identify and reach it (for instance the user's computer at work), it can present the flight selection on that device alternatively. The user can pick one of the flights and the agent will take care of the rest of the booking process, any payment issues, and of the reminder note on the user's PDA. To accomplish all these tasks, the agent will leave the user's computer again, use additional services somewhere else in the network, collaborate with other agents, and contact its principal in case of uncertainty, ambiguity, or confirmation.

No matter if it is an agent's home site, an arbitrary computer in the Internet that is simply used as a stepping stone on an agent's route, or one of the targeted servers, each device that should be accessible for mobile agents has to offer a standardized runtime environment. The main purpose of this environment is to provide the essential services and basic infrastructures that allow mobile agents to communicate, collaborate, and perform their computational work in general. Additionally, it offers standardized interfaces that enable access to external software systems. These interfaces can, for instance, be realized as Web Services or stationary agents which facilitate data exchange with databases, business applications or legacy systems. The runtime environment also provides mechanisms that allow mobile agents to freeze their execution state and migrate with their entire data to a new host. There, the agent can be reactivated and continue working. Such a runtime environment for mobile agents is called an *agency*. Connected agencies represent a logical network on top the physical one. This logical network represents the quasi-world for mobile agents and is called the *mobile agent system*. Finally, the (end-user) application, consisting of mobile agents which operate within the agent system, is called the *mobile agent application*. An agency can be seen as a

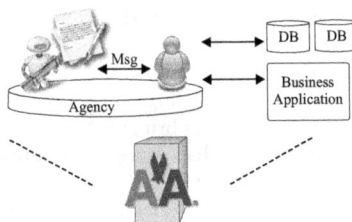

Figure 1.3: Interacting with host services, mediated by agency interfaces.

shielded container. Any interaction between a mobile agent and software components outside of the agency's boundary has to be mediated by certain interfaces, as it is shown in Fig. 1.3. Consequently, strong and effective security mechanisms can be employed to control agent activity. Especially, malicious agents can be blocked and terminated when they attack any service or the agency itself. Protecting agents against malicious execution environments, however, is far more difficult. Basically, agents are completely unprotected and must simply trust the environment they live in [Erf04].

If one thinks of the huge diversity of hard- and software systems, designing individual agencies for each constellation would cause extreme effort and immense costs, resulting in a high dependency between an agency and lower layers of the computer system, and making it difficult to exchange parts and components. To soften these drawbacks and to pave the way for agencies into the majority of computational devices, one often uses *virtual machines* (VM) like the Java Virtual Machine (JVM) [Lin99] or the Common Language Infrastructure (CLI) [ECM02] propagated my Microsoft. A VM covers any particularities of the host operating system and the underlying hardware, and represents an additional, standardized layer, making it much easier to deploy an agency. The JVM, for example, is available for almost every class of devices, from chip cards, to PDAs, to PCs, to mainframes. Of course one still has to adjust the agency's service interfaces, but with a VM one gets rid of a lot of issues concerning the operating systems and lower layers.

Comparing the web-based scenario in section 1.1.1 with the agent-based approach from above, important improvements become quickly apparent. First, in terms of time. The user can simply hand over the flight-related information to the agent. The time-consuming searching, browsing, and repetitive entering of data can be skipped completely. Once the agent has been given the itinerary, it handles the rest of the process autonomously. Second, besides time savings, the dynamic adaption of the agent's travel route makes it possible to discover and exploit new information sources that were previously completely unknown to the user. And third, the agent can be utilized as a means to sieve huge amounts of data and extract a surveyable fraction of relevant information already on the remote side. From a technical perspective, this can be restated as follows: in the web-based approach, entire web pages have to be transferred to the client where they can be filtered for information. Thereby, a star-shaped data flow between the client and the different servers becomes apparent, as it is visualized in Fig. 1.1. Agent technology pursues the opposite path. Agents serve as containers bringing algorithms directly to the server to filter data right on the spot. This has as least two advantages. Client devices can be kept simple, since most computational work is shifted to the server side. And network load can be reduced tremendously because only a slim agent is sent to a server where it filters relevant information and returns only with necessary results. Useless data does not have to be transferred and never leaves the server. This feature is known as code-shipping versus data-shipping [Bra05].

Although software agents can be considered as more or less regular computer programs, a fundamental difference remains in the way in which traditional programs and modern agent applications work. This difference boils down to the notions of computation and interaction and their evolution in the past fifty years of modern computing.

As outlined above, computation in the early years meant almost exclusively mathematical calculation. The following decades then re-defined this notion more and more generally as information processing with any kind of digital data—texts, images, audio, and video files. *Multimedia* was coined during that time and quickly became a buzzword. Today, mainly driven by the Internet, computation is essentially based on interaction. If one thinks of traditional programs, like spreadsheet applications or web browsers, they rely mainly on human-software interactions. Agent technology, however, shifts this focus decisively towards software-software interaction. Like the idea of clients and servers already represents a basic level of software-software interaction, computation in the sense of agent technology happens with a much higher degree of communication and social collaboration between loosely coupled and widespread software components [Luc03]. This conglomerate of interacting and collaborating software entities resembles virtual societies or even 'ecosystems' [Luc04] of software components.

1.1.3 Distinguishing agents from objects

As mentioned above, mobile agents at first glance resemble objects in the sense of object-oriented software. Both, agents and objects, are software components bundling algorithms and data, and both can be transferred via networks. But beyond these basic traits, comparing the two notions is like comparing apples and oranges. Agents and objects are notions situated on two different levels of abstraction. They can be distinguished from two perspectives.

First, with respect to their *objectives* of the technology. Object-oriented software design and the appropriate programming languages have once evolved out of the need for enhanced modularity and re-usability of code. These needs emerged out of the memory-saving and throughput-pushing programming in the previous decades, which resulted in fragile software systems and extremely high maintenance costs [Luc04]. Agent technology, however, does not address these objectives. Agents are faced with the challenges arising from the above sketched environment: interactive computational tasks between distributed software components and highly dynamic networks infrastructures. Instead of modularity and re-usability, the agent paradigm thus concentrates on adaptability and robustness.

Second, objects and agents are completely different with respect to their *technical nature*. Objects are always inseparable parts of an overall application. An object's life cycle is directly bound to the application's, and both mutually depend on each other. Objects are created when the application starts and they die as soon as the application shuts down. Furthermore, each individual object strictly operates within and according to an application-wide aim [Luc04]. That is, invoke an object's method and it will blindly obey. An agent, however, is an entity which is able to control its life cycle and its behavior autonomously. Although a single agent may be part of a multi-agent application and is given a clear objective, it will pursue its own strategy to achieve the intermediate steps towards the overall goal. This is

9

why the path towards the solution is much less determined than it would be when employing objects. Even when giving an agent the same task twice, it might each time use different information sources, services, and collaborating partners to fulfill the task. Agents must also be robust enough to deal with competing agents and changing conditions. An object finds itself always in the same environment and has only 'friendly' partners. These traits show why one cannot direct and invoke an agent as one can do with an object.

This autonomy can lead to situations where single agents persist longer than the overall application they are part of. Especially in case of mobile agents and their capability to migrate between different computers, this phenomenon becomes clear. They can keep working or simply wait on a remote server if their application on the home platform is temporarily shut down or becomes unavailable.

All in all, one can say that the actual technical link between the world of object orientation and (mobile) agents is that one can use objects to build agents. An additional, rather abstract link might be seen in the fact that agent technology is believed to facilitate the construction of complex, self-organizing, and social software systems, similar to how object-oriented ideas once led to system designs based on real-world objects. From that point of view, agents represent the new building blocks for the next generation of software systems [dS04]. These blocks possess the appropriate level of complexity and abstraction that upcoming applications will demand.

1.1.4 Promising fields of application

The fields of application for general software agents can be divided into three categories [Luc03]. First, this is the domain of simulations. Multi agent systems find broad audience when it comes to answer economical questions [Hic98, Pec98], to understand biological systems [Jon98], or to analyze phenomenons in human societies [Mag98]. No matter what specific simulation, agents normally represent real-life objects and interact according to an implied model. The emergent behavior of the entire system can then be analyzed, and gained results linked back to the real world.

The second field of application for software agents builds on the historical influence from research on Artificial Intelligence. Within this field, software agents are orchestrated to decision-making engines, analyze how decisions can be found, and how they can be re-transferred to concrete questions. These systems are of special interest to the economy [Sie04], but also find willing takers in the computer games industry [Meh03, Lew02].

The above mentioned flight booking example belongs to the third category. Researchers and applications within this field understand the role of software agents primarily as assistants, taking over certain responsibilities from their human users, acting on behalf of them, and assisting in information retrieval and selection [Kra97].

Mobile agents, as a special kind of software agent, fit best into the last category. Thanks to their mobility, they are able to convey algorithms to remote sites where data is produced or stored. Hence, they can analyze and filter data sets completely independently, and return to their origin with only relevant data portions. Mobile agent technology represents a new and more natural way to design software systems, depending on and retrieving information from remote data sources. This might be of special interest in situations where transferred data volume causes (monetary) costs.

Besides information retrieval, mobile agents can be employed as sensors at strategically important sites. If certain thresholds are exceeded, a specific information set becomes available, or a particular situation arises, those agents are activated and start working. Mobile agents are successfully utilized for the management and administration of computer networks [Rub99], monitor important devices in telecommunication infrastructures [Far00], assist in nuclear power plants, and can even be found in spacecrafts. They have conquered the domain of supply chain management [Huh02, Kum01], traffic control systems, the entertainment business, health care, and the military.

One of the most central points of application, however, represents mobile agent technology in the domain of e-commerce [He03]. Because of their autonomy and their ability to use services anywhere in a virtual market place, mobile agents purchase assets and use services similar how human users would go shopping in real stores [Bra05]. Mobile agents seem to be rightfully predestined to correspond to distributed workflows as they can be found in almost every business-to-business and business-to-customer application. Better then most other current technologies, mobile agents are able to handle error and conflict situations, evaluate alternatives, and recognize incompatibilities without any human intervention, and, thus, preventing the regular business process from disruptions. This great capability becomes clear in the situation when a targeted service is overloaded or temporarily unavailable, and the agent simply searches for an alternative service on some other host. The handling of these errors is completely hidden to the user. It is not relevant to him or her as to which location a service is used, only that it is used. This is one of central advantages of this technology, making it suitable for dynamically changing environments and superior to any centralized approach. A fixed assignment of the application's features with individual services (as used in a client-server system) would immediately fail if one of theses services became unavailable. A mobile agent, however, is able to find alternative services. So, every feature of an agent application is rather bound with a particular *service type* than a particular *service*. The mobile agents implementing these features can freely select between individual services of a type. That enables agent applications to deal with a steadily growing and continuously changing amount of services in the network, and guarantees much higher flexibility, efficiency, and usability which traditional approaches lack.

As colorful and diverse as the fields of application are for software agents, just so are the many agent toolkits which have been developed over the recent years. Many of them concentrate on particular aspects and are bound to academic research, but there are also some fully-fledged implementations and commercial products which have found their way into industry applications. The *Grasshopper* toolkit [Bau00] is one representative of the latter category. This toolkit is a compliant to both, MASIF [mas00] and FIPA [], the two driving standardization efforts within the agent community, and thus able to interact with many other agent systems. *ADK* from Tryllian [try06] has also reached maturity and was applied by Global IDs Inc. to implement a product suite, using mobile agents to monitor changes in enterprise systems.

From the historical perspective, IBM has to be mentioned, since they developed *Aglets*, one of the pioneers of agent toolkits [Lan98]. Security aspects are addressed and analyzed with *SeMoA*, a toolkit from the Fraunhofer Gesellschaft [IGD06]. *Tracy* [Bra04] finally, which has been developed at the University of Jena, and partially motivated this work, emphasizes the mobility aspect and allows implementation of fine-grained migration strategies for mobile agents.

For a comprehensive listing of projects in the field of agent-based computing, the following sources should be visited: A good overview on the European activities is provided on the AgentLink website (`www.agentlink.org`). IEEE Distributed Systems (`www.dsonline.computer.org`) maintains a list of major international projects.

1.1.5 Standardization efforts

As shown in the previous section, agent-based software systems can be found in various application areas. To coordinate the different projects, strengthen their interoperability, and provide a guiding framework, two organizations impel the standardization of agent technology. On the one hand, is the *Foundation for Intelligent Physical Agents* (FIPA) [fip04] whose standards cover the entire span of agent-based computing and concentrate on intelligent agents. On the other hand, the *Object Management Group* (OMG) publishes its MASIF standard [mas00] which is exclusively directed to mobile agents. Since 1999, both organizations collaborate and coordinate their activities.

The main objective of the FIPA standard is to enhance the interoperability between different agent toolkits. Therefore, various specifications within the standard are published, ranging from the core architecture of an agent platform to the support of semantic communication languages like FIPA-ACL/SL [Con00], KQML/KIF [Fin94] and the different interaction protocols. In the sense of FIPA, interoperability is mainly dependent on a common 'language' for agents. Such languages should be based on speech act theory, predicate logic, and ontologies. According to FIPA, interoperability can be reached if agents follow certain rules in communication and share a common understanding of their world [Luc03]. Most of current available agent toolkits follow the FIPA standard.

Whereas the FIPA efforts address agent-based computing in general, the MASIF standard considers mobile agent technology in particular. Systems, consisting of *intelligent* agents, gain their operational functionality mainly through message communication. On the contrary, a *mobile* agent system employs the mobility of the agents to realize application features. The features of such an application depend on mobile agents which migrate to remote services and use them. Therefore, MASIF mainly tries to standardize the different agent systems, e .g. the runtime environments. Interoperability in this sense then means that a mobile agent of one toolkit can migrate to an agent platform of any other toolkit. Any direct interactions between agents of different toolkits are not considered within the MASIF standard.

With regard to the challenges for mobile agent technology, as they are outlined in section 1.2, and this work's contribution towards a solution (see 1.3), one has to state that both FIPA and MASIF push the standardization mainly on the level of agents, agent applications, and rather isolated agent platforms. However, the infrastructural problems arising with dynamic large-scale agent systems are not addressed. Both standards assume a static agent system where services are closely bound to physical hosts and any dynamics are neglected. None of them explains how to setup and manage an infrastructure for realistic agent systems, and how to achieve robustness, efficiency, openness, and scalability.

1.1.6 Summary

Mobile agents are software components which encapsulate code and data, and resemble, at first glance, objects in the sense of object-oriented programming. However, the two concepts cannot be compared directly. They are located on two different levels of abstraction and pursue different aims.

Mobile agent technology represents insofar a new paradigm in software development, as the concept allows to describe and implement software systems, consisting of distributed, loosely-coupled, interacting, and 'nomadic' [Bra05] components. Their mobility enables these components to migrate autonomously within the network to remote services and data sources. They are able to separate relevant from irrelevant information right on spot, and thanks to their autonomy they don't rely on a permanent connection to their home platform. Thereby, they accommodate to current developments in the computer and network landscape, and score over many traditional approaches.

However, it has to be clearly said that mobile agent technology is not trying to blindly abandon conventional concepts for distributed systems. It is widely accepted that mobile agent technology won't breed any application that could not have been implemented using another suitable technology [Bra05]. This is one of the reasons why this technology has no so-called *killer application*. From a technical point of view, mobile agent technology only uses traditional concepts. The actual novelty has to be seen in the integrating force of the agent concept. It bundles various existing techniques into a consistent framework and thereby lends itself to providing the building blocks for the next generation of distributed software systems. It basically provides the adequate level of abstraction and complexity future software systems and their environment will demand [Bra05].

1.2 Vision and challenges

Using mobile agent technology induces a distributed system on two different levels. On the first level—the micro level—fine-grained structures are addressed. On this level, the agent application itself represents a distributed system. Even if it appears to be a consistent program to the end-user, from the technical perspective the agent application consists of individual components—the mobile agents—which might be distributed over the entire agent system. The agents use remote services to realize the features of the agent application. Using services means generally that a mobile agent interacts with stationary agents or other services provided by agencies [Dig03]. These stationary entities represent access points mobile agents can dock in to query a database or use any other system to get a job done. In either way, the mobile agent becomes a user of a service. This relation thus resembles the client-server idea. An important difference however is that the agent (client) can dynamically select a particular service (server) autonomously. In the classical understanding of client and server, the link between them is fixed and the client is much more dependent on a particular server.

Besides the interactions with stationary services, mobile agents can also interact directly. An agent may temporarily assist another, e.g. by sharing its query results, or both agents can arrange to collaborate over a longer period of time, e.g. splitting a task, solving the parts independently, and merging the results finally. Interacting in this manner, one cannot distinguish the roles of each participating agent. Depending on the situation, an agent can switch between both roles. This twofold character—being client and server—corresponds to the notion of a *peer* and the kind of interaction to the one found in *peer-to-peer systems* (see chapter 2). However, these interactions are not always altruistic. Agents always fight for limited resources, interfere, and can even hinder each other. Because of the similarity between these types of interaction and those observable in nature, this sight on the micro level inspired the term *ecosystem* which was referred to in section 1.1.2.

The agent application itself runs on top of the connected agencies—the agent system. This logical network represents the macro level of the above mentioned hierarchy. Of course, each agency is able to operate in isolation, but then agents are limited to employ only local services to get their job done. The features of such an application could be implemented much more efficiently with any conventional program. Thus, mobile agent applications make most sense if required resources and services are scattered among different physical devices. To make these devices accessible for mobile agents, they have to be connected to a network and each one must run an agency. From the agent's point of view, agencies represent an (almost always) equal environment, and thus, the logical network spanned by the agencies resembles a peer-to-peer system. It has to be emphasized that this is only true from the agent's perspective. See section 1.3 for further explanations.

Providing a runtime environment to the agents is the main purpose of an agency. It represents an additional software layer on top of the operating system and below the actual agent application. This is why such a layer is called *middleware*. The agency hides the heterogeneity of lower layers and provides standardized interfaces to make the host computer's services accessible to mobile agents. Hiding heterogeneity represents one of the two functions a middleware is responsible for. The second one lies in the management of resources. In general, the middleware handles any joining and leaving of resources and provides a consistent, seamless picture to higher levels of the software stack.

Here it becomes clear that the notion of the middleware has to be used carefully when referring to agent systems. Of course, a single agency can be seen as a middleware from the user's perspective. For him or her, it represents a consistent runtime environment for a specific agent application. It is hidden that there are other agencies and that individual agents of the application migrate to other platforms temporarily. For the user, there is actually no difference between a single agency and the entire agent system. Insofar, one can call the agency layer a middleware, since it hides any lower structures and gives a monolithic, coherent look onto the system.

From an agent's point of view this is only true as long as it interacts *locally* with one agency. Here, the agency covers the heterogeneity of accessing services on lower layers, and offers standardized interfaces to the agent. Resources, i. e. services, can simply be attached, removed, and exchanged unnoticeable to the agent. However, from the perspective of an agent migrating between different agencies, and thereby interacting *globally*, the agency layer cannot be called a middleware any more. The agent is fully aware of the distributed character of the agent system and any changes. In particular, the steady join and leave of agencies is noticed. That means the agent is fully aware of the system's distributed character. The single-system-image, as it is seen by the user, does not exist for mobile agents.

As already mentioned above, the agent system represents an inescapable virtual world for a mobile agent. To fulfill its user-given tasks, it is important to the agent to be able to find suitable services within this world. This essential necessity is emphasized again with the flight booking example in the next section.

1.2.1 A closer look at the scenario

Drawing on [Erf04] and looking again at the mobile agent in the flight booking example in section 1.1.2, the scenario might be restated as follows: At the beginning of the process, the agent is given a clear objective by the user. Here, it is at first the search for flight offers in the Internet. The user hands over the information necessary to describe the prospective flight. That is exactly the same information the user would use with search engines and on airline web pages. Out of the task description, the agent is then able to derive the services which are necessary to get this job done. On the basis of a virtual *map* of the agent system, the agent then tries to identify these services in the system. Similar to a human-readable map, agencies correspond to cities and network connections correspond to roads. The mapped services of the agent system could finally be compared to gas stations, restaurants, shopping malls, and any other service points indicated in real-life maps. These virtual maps are created and maintained on each agency.

The agent can use this map to spot individual services as well as the best network connections for its task. This information then leads to an optimal route for the agent's way through the computer network. According to this list of targeted services, the agent will start its journey and visits the selected agencies sequentially. From time to time, the agent should update its route, since some agencies on its list might have become unavailable while the migration processes. To avoid delays caused by downed agencies, and to be able to identify alternatives as soon as possible, the agent is always free to use the most recent map on any intermediate platform. Additionally, maps can be used to identify the best connection between individual agencies. If, for instance, the agent has already gathered a lot of data, it would be favorable to use broad connections for the rest of the journey. In other situations a secure connection would be preferred, especially when sensitive information is conveyed.

At the end of its journey, the agent returns to its principal and presents its flight selection. The latter picks one flight and then instructs the agent to book it. After being handed over the required information, e.g. the user's credit card number, the agent draws again on information of the agent system's map. This time, however, it has to identify services, assisting in accomplishing the booking process. Because the agent now carries the credit card number, the user name, and other confidential data, this is exactly a situation which demands for secure connections during migrations.

After finishing the whole process, the agent will return to the user's desktop, engage a stationary agent to print the flight confirmation, and put a reminder note in the user's PIM which is transferred to its PDA with the next sync process.

1.2.2 Analyzing the scenario

Throughout the scenario it becomes clear that one challenge is of major importance to the whole concept: How does a mobile agent navigate within the waste amount of services in the agent system? This problem can be divided into three aspects:

- The agent must be able to understand the user-given task.

- It must be able to derive the appropriate types of service out of the task description, identify their interdependencies, and arrange them accordingly.

- Finally, it must be able to assign individual services to the identified service types.

The first two points are clearly the most difficult. Not exclusively from agent technology's point of view one has to admit that machine-based comprehension of human-given tasks requires a high level of maturity and an even higher degree of intelligence of any computer system. The agent has to capture the essence of the task and therefore understand the meaning of the given formulations—thus their semantics. For the second point, the agent must first bridge the gap between the human semantics of the task and the artificial semantics of the software services. Then, the agent has to identify dependencies between these services and align them appropriately. The last point finally seems a lot easier compared to the others. There, the agent 'simply' has to fill the blueprint built-up in the first steps with concrete services offered somewhere in the agent system. Thus, point three represents a mapping process where individual services are assigned to the service type slots. In other words, the agent simply has to decide whether to use a service on agency A or on agency B.

It has to be clearly stated that for the first two aspects there exist only partial approaches, and both areas still represent very vital research fields. Although ontologies provide a meaningful concept to categorize service descriptions, they are neither able to efficiently cover every imaginable service category, nor they can be easily compared or merged in the majority of cases. Other approaches try to induce 'intelligence' into software systems by making use of neural networks. However, the largeness of the required data sets and the complexity of such systems limit this idea to applications using rather intelligent agents than mobile ones. Using a neural network directly within a mobile agent would heavily inflate the code size. That in turn would contradict one of the major advantages of mobile agent technology which is—simplified spoken—sending a thin agent to a huge data basis to filter it. The 'intelligent mobile agent' would become so big (in terms of code size) that downloading its targeted remote data with a conventional client-server mechanism would become quickly more efficient.

A very promising approach is introduced in [Gle04]. It combines both types of agents and tries to avoid the ambiguities arising with point one of the list above. First, intelligent agents, embodied as *avatars*, are utilized to interrogate the user. They try to capture the user's demand as precisely as possible by posing well defined questions and thereby shrink the space for speculations and improve clarity. The derived task description can then be handed over to a mobile agent which continues working as depicted above. With this approach the complex and large logic for the intelligent agents can be kept locally and the mobility of mobile agents is not weakened

However, even this approach cannot be applied in every situation; the complexity of the first aspects is too high. Since currently no approach is ready to be used for mobile agent environments, the first two points are not further regarded in detail. Their general feasibility, however, is not doubted and they are taken for granted hereinafter.

This book focuses on the third aspect of the list. The matching between a specific service type and a particular service does not require any additional semantic interpretation in general. The agent simply has to select one individual service out of a set of equal and interchangeable services of the same type. However, even this selection should not be done indifferently. Being aware of the fact that services are always bound to agencies, and that some agencies are more distant than others, proximate agencies should be preferred if fast answers are expected from an agent. If it is foreseeable that an agent must gather large amounts of data, agencies with broad network connections should be selected. Much in the same way other criteria can easily be found which make some agencies always appear better suited than others.

According to [Erf04], agent toolkits can be separated into two categories, depending on how mature they are already with respect to the third aspect to the third point:

1. Agent systems of the first maturity level.

   a) The mobile agent itself, its objective, its strategy, and its route are entirely determined during its implementation. That means that the agent follows a 'hard-wired' travel route through the network. If services on that route become unavailable and network connections break, the agent will no longer function and has to be re-programmed by replacing the obsolete services and routes in the source code.

   b) The mobile agent itself, its objective, and its strategy are defined during its implementation. However, its route (i. e. the individual services) is determined by the user. In contrast to a), here only service types and their arrangement are fixed. The concrete services can be assigned at runtime; hence the agent needs no re-programming. The agent is given the route, like a conventional program is invoked with parameters.

2. Agent systems of the second maturity level.
   A mobile agent of the second level is almost as equally determined as an agent of the first level. Its basic skeleton, its purpose, its strategy, and its list of service types are all hard-coded. The only—but crucial—difference to a), where the route is hard-coded, and b), where the route is passed as parameter, is that an agent of the second maturity level is able to select suitable services autonomously and without human intervention. Such agents are also able to adjust their service list dynamically in the course of their work. If a service is found to be unavailable, the agent simply replaces the broken one with an equivalent service of another agency. These capabilities have been referred to in section 1.1.2, using the terms autonomy and intelligence. In the sense of this work, autonomy and intelligence solely mean the capability of a mobile agent on the second level of maturity to intelligently and autonomously select suitable services out of the myriads of service descriptions within the agent system. The emergent agent behavior is then called service-oriented navigation [Erf04].

The outlined scenario in section 1.2.1 thus requires an agent system of the second level. For a mobile agent, two things must be assured in order to efficiently select and assign individual services: First, the agent must have access to some sort of database containing the service descriptions of the agent system. As presented in

[Erf04], this database is called *map*, since it gives a logical view on the agent system, similar how a real map can be used to get an overview on an area. The second essential requirement for an efficient selection process is determined by facts like how accurate, how far-reaching, and how old the map information is. Especially the latter attribute plays a crucial role.

If the agent system consists of only one agency, all these requirements can easily be granted. The descriptions of the services provided by the agency can be published locally, and thereby become accessible for all agents. Tracy (see section 1.1.4), for example, can setup and maintain a so-called *blackboard* which can be used by agents to search for locally advertised services and information sources.

However, the larger the agent system gets, the more desired qualities of the map begin to contradict. For example, real-time capturing of the available services becomes more and more difficult the larger the system grows. If each agency would try to steadily capture the dynamics in the system's service availability, bandwidth of the physical network would quickly become a scare resource. *Jade* [Bel01b, Bel01a] tries to overcome this problem by running a central service directory on one particular agency. All services of the agent system must register at this directory and thereby become published and accessible to agents. Like always however, a centralized structure is highly fragile in case of important components break down. If Jade's directory does not work properly, the entire agent system is forced to suspend almost all activities. Additionally, the Jade agency, running the directory service, must handle a much higher work load, since all requests for (de-)registration of services as well as all queries of the agents involve the single service directory.

Finally, in large-scale agent systems, it becomes impossible to fulfill all claims regarding the quality of the map information. Any centralized approach for the management of the service information (like Jade) would overload single agencies sooner or later, and any straightforward decentralized approach would bring the system down to its knees too because the traffic would jam the network. In the best case, such a system would scatter into clusters when the weakest connections break. Each cluster then is either a small group of agencies with little maintenance overhead, or a group with broad interconnection, capable of handling a larger overhead. However, in either case each cluster would be an isolated, closed mini agent system that mobile agents can't escape from and thus can't reach services in other clusters.

If is seems to be impossible to manage the mapped information on each agency, one could argue in favor of equipping each agent with its own map. However, this too seems to be impracticable, since first, the map itself and the necessary algorithms would enlarge the agent in terms of code size, and second, each agent would become personally responsible to maintain its map. The former fact would imply that, depending on the agent's task, different information needs to be captured in the map, and consequently, particular algorithms would be required. Since it can't be foreseen which tasks have to be accomplished by the agent during its lifetime (even if all jobs certainly fall to the agent's general expertise), it would become likely that the algorithms need re-work because other information is required for some job. The latter aspect of a map-equipped agent means not only that the agent is forced to spend less time with its actual task, but also that each agent would cause network traffic while trying to inquiry the available services on remote agencies. All in all, equipping each agent with its own map seems to be less efficient, and maintaining a map on the agency is the only way out. This map should contain general-purpose information on all services in the system and not exclusively those required for a particular task. This allows all agents to benefit of this map [Erf04]. Furthermore, maintaining only one map on the agency seems to be more efficient with regard to network load than forcing each agent to maintain its own map.

However, the problem remains that with growing size of the agent system it becomes more and more impossible to capture the service availability of the entire agent system on a fine-grained level and in real-time. As mentioned above, if only one agency maintains one central map for the entire agent system (similar to the Jade approach) this architecture can't scale with arbitrary large agent systems, and the central agency always represents a single-point-of-failure. On the other hand, a naive decentralized approach, where every single agency tries to capture the status of the entire agent system on a map, must fail, too.

It can be summarized that maps should always be maintained on agencies. Neither a centralized nor a naive decentralized approach allows gathering of all desired map information in an efficient manner. To overcome this dilemma, trade-offs must be made regarding the individual qualities of the map information.

Erfurth demonstrates in [Erf04] just such a trade-off. He uses the agent toolkit Tracy [Bra04] that distinguishes between three different roles an agency can occupy. The first role is the one of a *regular agency*, a standardized runtime environment for mobile agents as generally referred to within the previous sections. The second role a Tracy agency can play is the one of a so-called *domain manager*. Similar to how computers are connected to a sub-network by a router, a domain manager groups several agencies to a logical organization—a Tracy domain. All services within such a domain are registered, managed, and published via the domain manager. Each agency joining the domain registers with the domain manager, becomes integrated, and can publish its services by a directory service running on the domain manager. Depending on the utilized strategy, two ways of searching for services within a Tracy domain are possible.
In the first case, the domain manager copies the most recent directory content to each agency of its domain whenever this becomes necessary, e.g. after adding a new service or deleting an old one. An agent can then simply use this local copy to search for services within the domain. In the second case, the domain manager does not share the directory content with the associated agencies. Consequently, all agents of the domain have to query the domain manager to find a service. This strategy is similar to the one pursued in Jade with its so-called *Yellow Pages Service*. In either

way, both strategies in Tracy allow an efficient search within a domain as long as
its size is restricted.

The third role known in Tracy is the one of the so-called *master*. This role is oc-
cupied by exactly one agency in the whole agent system. The master represents
the root of the virtual tree which is induced by the roles within a Tracy agent system, and interlinks the domain managers of the individual domains, as it is shown
in Fig. 1.4. Analogous to a domain manager, the master maintains a directory

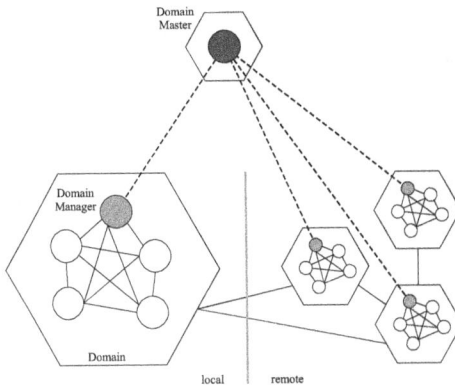

Figure 1.4: The Tracy domain concept. (Drawn on [Bra04]).

which is used to publish services on a global basis. Domain managers can rely on
this global directory if an agent's request for a particular service can't be satisfied
within its own domain. The global directory will contain a record of that service
if it has been published in some remote domain. Thus, even services in distant
domains can be found, and a mobile agent can leave its current domain to migrate
to a new one.

However, to reduce the amount of data in the master directory, domain managers
submit only a coarse-grained profile of their domain. For instance, if a service can
be found on several agencies in one domain, the domain manager would only publish *that* the service is available in its domain, but not on which particular agencies
exactly. An agent coming from another domain is thus forced to query the local
domain manager again to get the references to the individual agencies offering the
service it is looking for.

Based on this idea, Erfurth suggests that the maps maintained on the domain
managers (and which are then eventually copied to the attached agencies) could
benefit from this data reduction enormously. This motivated him to introduce so-
called *fish-eye-view* maps. Instead of trying to capture the services of the entire
agent system, each domain manager primarily focuses on its own domain first and
depicts the locally found services in detail on the map. This portion of the map

then corresponds to the sharp-focussed center of a fish-eye. Then, the domain manager can draw on the globally published data in the master's directory. With that information the second portion of the map is built. It represents the blurry border of a fish-eye where things are less recognizable. The reduced profiles of remote agencies, as they are stored on the master, serve to extend the local domain map at its borders. Such a fish-eye-view map then gives a limited view on the entire agent system, with fine-grained information in its center and increasingly coarse-grained information towards its boundaries. Merging all maps of the individual domain managers would then give a coherent snap-shot of the entire agent system. Because of these scattered pieces of information, the fish-eye-view concept follows a distributed approach. However, compared to the above-mentioned naive idea, it represents a clearly advanced technique. With a fish-eye-view map, all services of the local domain can be found immediately by any mobile agent residing in the domain. Remote services can also be identified within two steps: the first, when querying the local map which then refers to a distant domain, and the second, when querying the map maintained by the remote domain manager to finally identify the service within its domain.

The fish-eye-view concept leads to a massive reduction of the overhead that arises with the creation and maintenance of the maps, while still guaranteeing a powerful enough search mechanism for service discovery. However, this concept, too, is not fully free of weak points. One of the most severe ones has to be seen in the infrastructural role of the master. If this central agency breaks down, the entire agent system scatters into separate fragments and thereby shrinks the radius of action for the affected mobile agents.

In brief one can summarize: The majority of current available agent toolkits is still settled on the first maturity level [Erf04]. Routes are explicitly determined already during the implementation of the mobile agent, or they are passed as parameters when starting the agent. This fixed coupling contradicts the expected dynamics within a large-scale agent system and disqualifies any toolkit of the first level for autonomous, service-oriented migration of mobile agents. Erfurth shows that autonomous navigation can be realized, using virtual maps of the agent system. In particular, he proposes fish-eye-view maps which facilitate both local and global searching for services, but keeping the maintenance overhead and network load at a low level. However, the hierarchically structured agent network he uses to introduce his concept contradicts the decentralized nature of the agent system and makes single-point-failures inevitable. This drawback is addressed in this book as outlined in the next section.

1.3 Contribution of this work

Erfurth's concept of the so-called fish-eye-view maps, as it is sketched in section 1.2.2, represents an excellent means to search for services in the entire agent system while keeping the volume of overhead data small. Hence, fish-eye-view maps are a reasonable trade-off between the most important, but in large-scale agent systems contradicting, requirements for service-oriented navigation. Worth highlighting again is the compromise fish-eye-view maps find between the captured information's level of detail, and the gained quality of search.

The hierarchic structure of the agent system depicted in Erfurth's concept is induced by assigning different roles to the agencies. If essential entities of this infrastructure become unavailable, namely the master agency, the global life environment for the mobile agents splinters into isolated domains, making it impossible to create the proposed maps, and ultimately prevent the agents from efficiently fulfilling their task.

This book regards the concept of fish-eye-view maps as an efficient means to enable service-oriented navigation within an agent system. However, a hierarchically structured agent system with clearly assigned roles of the agencies is considered less beneficial. Not only that it contradicts the inherent equality of agencies, but it also forces the whole functioning of the agent system on a few central entities. In addition to the lack of robustness, the centralized traits of Erfurth's exemplified agent system do not scale with an arbitrary large and dynamic agent system.

With this background one can derive an optimal environment for mobile agents and their demand for autonomous service-oriented migration:

- The infrastructure of the agent system should strictly follow a distributed approach. Failures of single agencies must not lead to a break-off of the entire system. Furthermore it must not cause that services in the agent system cannot be discovered anymore. Every agency must be able to occupy any role within the system, so that in case of an agency's break down, its responsibility for a certain system-relevant feature can be transferred to another agency immediately.

- Any logical structure, hierarchies in the community of agencies, distinctive roles, etc., should be completely hidden to the agents. Furthermore, any dynamic adjustments within the infrastructure (as mentioned in the previous point) should be transparent to the agents, too. Instead, they should always find an equal environment on any agency, making any interaction as standardized as possible.

- Because of the dynamics to be expected in the agent system (joining and leaving of agencies plus changing availability of services), (un-)mounting of platforms and their services must be done close to time to allow agents engaging assets which are only available for a very short period of time.

- The management overhead arising with the growing size of the agent system must not influence the efficiency of the agent's work. Especially the requirements for the virtual maps must not be hurt and administrative network traffic should be as low as possible.

To comply with all these points, this book emphasizes the agency's peer character as it was revealed in section 1.2. In the envisioned setup of the agent system, one should not see any roles or hierarchies, and all agencies should be interchangeable with respect to their position in the entire agent system. This strict equality leads to an infrastructure where no agencies are of central importance to the whole system. Thus, all single-point-of-failures and hot-spots for malicious attacks can be removed, and the system would gain a much higher level of robustness.

The realization of the first two points then induces an agent system, coming close to the orchestration of a classical peer-to-peer system. However, this comparison is only partially correct. Exactly which traits are different is presented in chapter 4. Also in chapter 4 is a description of how to deal with the flood of information and the resulting management overhead from the system's dynamics, as it is mentioned in the latter points. Finally it is shown how to build the fish-eye-view maps within the new structure of the agent system. The next chapter, however, first introduces central characteristics of the peer-to-peer technology, as they are required for the understanding of the following sections.

2 Peer-to-Peer technology

The client-server architecture, referred to in section 1.1.1 as the dominant paradigm for Internet applications, slightly covers the fact that the original architecture of the Internet does not know such separations of roles. The original purpose of the Internet was to be a communication infrastructure for message exchange between independent computers [Lös05]. Within this network, each member/computer is equal in the sense of a peer, and hence the network resembles a peer-to-peer system. *Usenet News* probably represents best this original intention.

Tracing back the Internet's history even further in time, one gets to the *ARPANET* (see 1.1.1). The development of this predecessor of the current Internet was primarily motivated by the Cold War and the fear for an intercontinental nuclear strike. With *ARPANET* one tried to establish a decentralized and fault-tolerant computer network that could stand such scenarios. Therefore, independent computers were interconnected and could be used to exchange data. If one node broke down the traffic could be routed via other computers without fear of network disruption. Thus, the whole system was able to keep operating regardless of individual node failures.

In the following years, however, peer-to-peer systems did not find popularity on the application level that client-server programs did. It took until the mid-1990s before the concept experienced a revival with the booming Internet. Instant messaging systems, file-sharing applications, and clients for distributed computing suddenly sprouted-up everywhere. *Skype* (www.skype.com), *Gnutella* [gnu04], and *SETI@home* (http://setiathome.berkeley.edu) are some popular examples. With Google's *Desktop Search* and similar applications facilitating the enterprise-wide search for arbitrary contents from any employee's PC, peer-to-peer technology has also found its way into the business environment. Even the academic sector has got its specialized peer-to-peer applications. Bibster [ea04], for example, discovers and shares information on published papers within the research community.

Fig. 2.1 shows some well-known peer-to-peer systems and classifies them according to their field of application [Bre01]. Most of the contemporary peer-to-peer applications pursue one of the following major goals: They either harness multiple desktop computers to become a virtual supercomputer, they network distributed information resources, or they facilitate collaboration between individuals beyond organizational and geographical borders.

In the first instance, a complex problem usually is divided into multiple sub-problems which are then distributed to numerous computers via a peer-to-peer network. All computer simultaneously solve their given sub-problem and submit the result to a central server where they are merged to form a coherent solution for

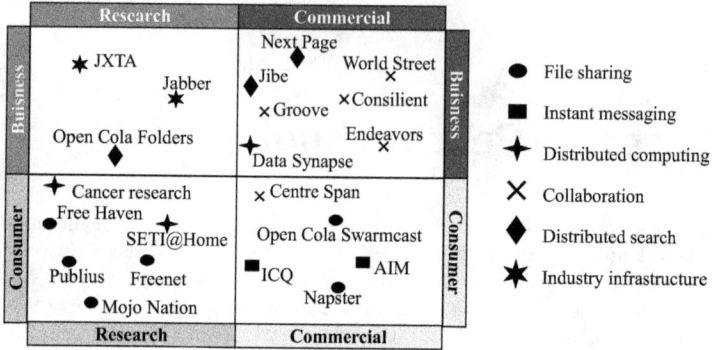

Figure 2.1: Popular peer-to-peer system and their domain of application.

the initial problem. The SETI@home project, for example, can relay on a computational power of 25 teraflops to search for extraterrestrial radio signals by pursuing a divide-and-conqueror strategy on the back of a peer-to-peer network.

One can find similar approaches in commercial environments, too. Complex aerodynamic analysis, crash simulations, gene sequences, molecular structures, and seismic predictions are computed on networked computers connected via a peer-to-peer system. However, in the pure sense these virtual supercomputers can't be termed peer-to-peer systems because the individual parts rarely interact with each other— if ever. The infrastructure of these systems is more comparable to a pull-push client-server approach and falls in the domain of grid computing [Var05] and related technologies. But even if the term is slightly misused, it shows that the idea of peer-to-peer technology also finds its audience in business environments and that valuable inspiration for future developments can be expected from this area.

Linking scattered information, distributing content, and sharing files characterize the second field for current peer-to-peer employment. Besides the myriad of file sharing programs for private use, this idea can also be found in business applications. Today most files within a company aren't actually stored on central servers but on the employee PCs [Bre01]. Peer-to-peer systems connect these contents to a distributed file system every employee can use to find a desired document. Besides this pull-mode, many of these systems can also be used in a push-mode where individual files from a central server are uploaded to the employee PCs. Intel Corporation, for example, uses a peer-to-peer application to distribute digital training material among its employees[1]. This software is also able to capture changes in its neighborhood and thereby offers caching capabilities. If, for instance, one member of a workgroup has already received or downloaded a file from a central source,

---

[1]Microprocessor Research Laboratory Beaverton, Oregon, USA

the peer-to-peer application on each computer will download the content from that one employee's PC instead from the server. Thanks to this kind of caching, downloading files is much faster, employees save time for more important issues, and the workload of central servers can be reduced significantly.

When it comes to collaboration, even beyond organizational and geographical borders, peer-to-peer applications connect people and create a real-time knowledge sharing system. There is no need to setup central servers or security guidelines in advance. A distributed workgroup can start working immediately and is much more flexible. Without loosing time for administrative issues, the participants connect to the system ad-hoc and share resources as if they were sitting next to each other. *Groove* (`www.groove.net`), for example, a peer-to-peer based working environment, allows online and offline work. In the latter situation, changes are logged and cached locally and synchronized with the other members of the workgroup as soon as the computer gets online again.

In brief, the concept of peer-to-peer based interaction can be found on various levels of abstraction: Beginning on the network layer (Internet), to the information management layer (e.g. distributed database systems like Mariposa [Sto96], to the application layer (e.g. e-commerce systems like eBay), and finally even to the top layer of human users (e.g. eBay users interact much like peers within a virtual market place).

With the advancing spread of peer-to-peer systems (on the application layer), allocation and exchange of distributed assets is enhanced and accelerated dramatically. Peer-to-peer systems are believed to present an alternative and very beneficial concept for future software systems since they mutually correlate with the characteristics of the working environments they aim at. The clear separation of roles between asset-providing servers and asset-consuming clients, as it widely found today, is replaced by a system of peers. With equal rights and responsibilities they form a mesh that allows ad-hoc linkage of resources, instant content sharing, collaboration, and even possess self-organizing and self-healing qualities [Lös05].

2.1 Characterization and classification

As seen in the previous section, a lot of currently available software claims to have peer-to-peer qualities. But on closer inspection, it is sometimes nothing more than a marketing campaign to push business. To be actually called peer-to-peer, a system must have the following properties [AT04]:

- Symmetry of roles – Each peer of a peer-to-peer system can serve as both service supplier and service taker, and can switch dynamically between the roles of server and client. One also finds that a peer is termed *Servent* [Ste04]. With the equality gained, each peer also takes over certain responsibilities for the functioning of the overall system. Usually, this means that a peer offers a certain amount of its storage space and CPU cycles to other peers and/or handles a portion of the overhead to maintain the system structure.

- Fully distributed – In pure peer-to-peer systems there is no central coordination at all. In particular, each peer's sight only reaches its own neighborhood. At no time it is possible for an individual peer to gain a comprehensive view of the entire system. Furthermore, these systems do not possess a central data base. Instead, each peer stores a fraction of the available data and ensures visibility to the other peers. A fully distributed design avoids single-point-of-failures and enhances the system's scalability and robustness [Mil04]. Thereby, not only workload and responsibilities can be distributed among different computational entities, it also represents a powerful means to prevent ideological threats, like sabotage, spying, and censorship.

- Self-organization – The total distribution of information and knowledge in a peer-to-peer system and the lack of any central control authority forces each peer to be highly independent; although this might be seen as a hen-egg relation, too. Basically, a peer is autonomous and can survive without the system, but then it will usually loose almost all of its functionality. The qualities, behaviors, and structures of the whole system emerge quasi unintended out of the individual interactions from the peer level.

- Local interactions – As already touched-on in the previous point, the organization of the whole system is based on local interaction between peers. Of course this also comprises integration and removal of peers. Anything can be handled on a local basis without central control. The system's structure dynamically changes as a result of these modifications.

- Fault tolerance – Although lacking of central control mechanisms, and imposed to completely distributed information, every peer in the system must be able to find all published contents at any time. The system cannot be restricted by interruptions resulting from the down time of peers or a break in network connection. Suitable countermeasures, such as replication and other types of redundancy, should be in place to deal with such occurrences.

However, as already seen in the previous section, seldom are all criteria implemented to the same degree. The majority of implementations violate the infrastructural aspects of the first points. This leads to the following classification where peer-to-peer systems are divided according to their basic peer infrastructure [Mil04]:

- Fully distributed systems – Such systems comply entirely with the infrastructural requirements mentioned above. All peers are completely equal and communicate directly without dependence on intermediate peers. Gnutella (till v0.4), Freenet, CAN, CHORD, Pastry, and P-Grid fall into this category.

- Super-peer systems – Some peers of a peer-to-peer system are likely to be more powerful than others. With broad network connections and/or greater computational power they are able to take over major parts of the maintenance and administration overhead arising within the system. They can thus reduce workload for weaker peers [Gon01]. Such powerful peers are called *super-peers* because they are raised within the community. Normally they stay constantly connected to the system and serve as entry points for other peers. Thus, each super-peer develops an associated group of clients. Search requests as well as all other communication from that group are mediated by the super-peer which takes care of the correct routing and publishing etc. of messages. The super-peers, in turn, are connected to a pure peer-to-peer structure and facilitate all interactions between the individual groups. This concept induces a two-level hierarchy in the peer-to-peer system where super-peers represent the top level and build the backbone of the infrastructure. Edutella, for instance, is such a super-peer-based implementation [Nej04].

- Hybrid systems – This class of systems is influenced by both architectural concepts: client-server and peer-to-peer. Implementations of this category always possess a central server in their infrastructure. Such servers administer the system and serve as unique authentication and entry points for peers. They also maintain a global index of all published content in the system. Hence, any search request is routed first to the server which replies with the list of peers storing the requested content. Based on this list, the caller peer can then establish direct communication paths to the referenced peers and retrieve the content. Hybrid systems can always be identified by this twofold type of communication—at first a classic client-server request to retrieve information from the global index, and then direct peer-like interactions with other nodes. Skype, for example, uses a central server to authenticate users and temporarily cache messages. And Napster [Nap03], once one of the pioneers of this field, was quickly forced to its knees by shutting down the central server after severe copyright violations had occurred.

The above listed characteristics of peer-to-peer systems represent advantages as they would be clearly impossible with classical client-server approaches. Enhanced fault tolerance and scalability are just two of the gained strengths. However, on the other side of the coin, these advantages result in much higher complexity and augmented administration efforts. The basic reason for this is that peers, other than clients and server, are not directly connected. Interaction between two peers in a peer-to-peer system is often mediated by other peers. Therefore, similar how content is conveyed in the Internet, routing mechanisms must be present in such systems that facilitate communication between individual peers. In hybrid systems, central servers negotiate between peers, and in super-peer-based as well as in fully distributed implementations, messages are transported by a kind of multi-hop routing mechanism. Because of this infrastructure, peer-to-peer systems are challenged to provide algorithms well adapted to the distributed environment.

Although various solutions for important functionalities have been studied in detail and are already available, some areas still need a lot of additional research. Transactions, for example, a well functioning mechanism in many client-server applications, are virtually impossible to implement in current peer-to-peer systems [Hau05b]. The next section addresses an additional challenging field within research on peer-to-peer systems: How to discover published content in such systems?

2.2 Information retrieval

Peer-to-peer applications generally possess a huge number of peers. Due to this size, the dynamics of peers and/or their content, and the lack of any central coordination (at least in pure peer-to-peer systems) information discovery and retrieval becomes one of the top challenges for such systems. How efficiently information lookup can be realized depends mainly on the system's basic structure. In addition to this general nature, classified accordingly to the basic peer layout (see 2.1), the way how content is published leads to a combined structure of peers *and* data. This logical structure is termed *overlay network* [Abe05], since it overlays the physical network. Thus, it can be restated that the efficiency of information lookup and retrieval depends mainly on the structure of the overlay. One introduces the following classification [Lös05]:

- Unstructured without indices – Peers and contents are randomly integrated into the system. There is also no indexing of published items. Hence, searching is often realized as a breadth- or depth-first search by blindly flooding the system with search messages. Even though one can limit a message's time-to-live (in terms of hops), this strategy obviously causes high network load. Furthermore, since one cannot predict the set of peers the messages will reach, there is no guarantee that requested content is actually discovered, even though it might be available at some node(s) in the system. Additionally, the system's search response time is quite low. Although one cannot overcome these drawbacks completely since they are intrinsically bound to the system's inner structure, there are several approaches to reduce their impact. Replication, for instance, and so-called *Random Walks* [ZY04], where search requests are sent out to a (smaller) group of randomly selected peers, can shrink message volumes and improve response times. Gnutella v0.4 and FreeHeaven, for example, are implementations with unstructured overlays and no indexing strategy.

- Semi-structured with local indices – Peers of such systems maintain a local index in which they store information on published items and other peers. Over time, each peer fills and updates its index based on information gathered from interaction with remote peers. E.g. two peers can exchange references stored in each other's index when they meet. The more references a peer can acquire, the better it becomes integrated into the system, and generally this increases the quality of search, since requests can be routed towards promising sources. Freenet, just to name one example, uses such an adaptive routing scheme where all search requests and responses are analyzed by each node they pass. The collected information is then used to build-up a peer's local index and thereby captures constantly better the system's structure as well as the availability and distribution of contents.

- Structured with global indices – In contrast to unstructured implementations in which peers and items are added randomly to the system, structured systems have a set of explicit rules which strictly define all aspects of integration. Usually, unique identifiers are assigned to relevant entities like peers and items. Hash functions are often used to derive such keys. The mapping of keys and their associated entities is then stored in a global index which is distributed among the peer community—a so-called *distributed hash table* (DHT). Than means each peer stores only a small piece of this global data structure and some additional links to other pieces. The identifier finally determines at which position in the system an entity has to be integrated. Searching is then realized by firstly deriving the identifier from a search request by applying the hash function. This will generate a key that matches the one of the requested item. Its position can be located by querying the local index. If it does not contain the reference directly, it contains the information about which peer stores the appropriate portion of the DHT that contains it. In the latter case, the request is forwarded to the identified peer. Whether only peers or peers and items are indexed, one distinguishes two sub-categories:
    - Peers only – Such systems assign unique identifiers only to peers and mount them at the corresponding position in the system. Structella, for example, integrates peers to a virtual ring topology which then allows efficient broadcasts. Similarly, Edutella uses a HyperCup topology.
    - Peers and items – Here, both peers and contents are assigned unique keys and placed at the corresponding position in the system. As soon as the keys are published within the overlay, they represent links to the associated contents. Based on the derived identifier, a search request is then forwarded to exactly that peer which is responsible for the corresponding part of the global DHT referring to the requested content.

According to [Kel02], unstructured and semi-structured implementations are well suited if the following properties apply:

- Peers demonstrate high dynamics. They connect often and stay connected for rather short periods.
- Mainly popular items stored on several peers are published in the system.
- It is acceptable if not all items are found matching a particular search request. It is sufficient to discover an incomplete list of the most popular or similar items.
- Search is based on key words.

Structured peer-to-peer systems, however, fit better if the following attributes describe the environment [Kel02]:

- Scalability is the core requirement. Large numbers of peers and items are expected to be administered.

- Peers and contents are rather static than volatile.

- Most items are not popular and are stored on only a few peers.

- Finding similar items is not sufficient to satisfy a search request. An exact match between the search term and the item is required.

In chapter 3, selected peer-to-peer systems are presented in greater detail. It is shown which overlay structures they possess and how they are exploited to access stored information.

2.3 Self-organization

As already mentioned above, pure peer-to-peer systems do not have any central authority that coordinates the activities of individual peers. Interaction occurs only on a local basis, mainly between adjacent peers. Combining all these local actions result in the global system's behavior. On this level, properties emerge and new qualities become apparent that do not exist on the level of individual peers. In particular, the phenomenon of *self-organization* is of central interest. If a system possesses self-organizing qualities, it is able to adapt and maintain its structure according to the environment it exists in without any interference from outside. In case of a peer-to-peer system, self-organization means that the system adapts to the churn of peers and contents without human intervention. Furthermore, the self-organizational capabilities of a system lead to an enhanced robustness, higher elasticity, and even self-healing qualities. Gnutella was one of the platforms to study this phenomenon.

Gnutella's interaction protocol comprises two parts. The network maintenance protocol on one hand, is required to build and maintain the overlay. The search protocol on the other hand, is used to discover and retrieve information out of the overlay. Initially, a peer uses the maintenance protocol to discover already integrated peers in the Gnutella network. It sends out a flood of ping-like messages that are answered with a special reply if they reach a Gnutella peer. From the replies received, the peer then selects several remote peers, establishes direct network links to them, and thus becomes part of the overlay network. The number of direct network links is usually limited. If broken links are found, because targeted peers might be down, the peer tries to establish alternative connections to other peers. The structure of the overlay network thus changes steadily and adapts to new constellations within the peer population. This represents exactly the process of self-organization. The force behind this process is the 'noise-driven variation' [Sta03]. Noise represents the 'fuel' supplying the system with energy and driving it into the different regions of a state space. In the case of a peer-to-peer system, noise results from the constant joining and leaving of peers, their random interactions, the resulting links, and from any time-related delays in peer communication. The state space is defined by all possible layouts of the overlay network. The system will then most likely strive towards the dynamic equilibrium of a steadily changing overlay.

The resulting overlay in Gnutella has two characteristic properties which can be found in a similar way in many other peer-to-peer systems. On the hand it has a small diameter, i.e. two arbitrary peers are connected by a rather short path in the overlay network. Thanks to this characteristic, search requests can be quickly distributed and reach promising peers even though the peer that sent out the request does not 'know' about this positive effect on the global level. It simply floods the system with messages and 'hopes' that they reach as far as possible and find a suitable peer. The second property observed with the Gnutella overlay is that the network's node degrees follow a power-law distribution, i.e. only a few peers possess a high number of links whereas most others possess significantly fewer. This fact is based primarily on Gnutella's maintenance protocol. While trying to discover already integrated peers of the system, joining peers will discover rather static peers more easily than those which connect only for short times. Hence, static peers evolve to central access points, become more and more connected, and build the system's backbone [Bar99].

A small diameter and a power-law distribution of the node degrees are exactly some the characteristics of so-called *Small World* graphs [Piv05]. They are often the product of self-organizing processes and can be characterized by the following attributes [Lös05]:

- Small diameter – The distance, measured in hops, between two arbitrarily chosen peers of the system is astonishingly small. In general less than eight hops.

- Sparse connectivity – Compared to a fully connected network, where each node is connected to all others, peers in Small World systems have usually very few connections.

- Power-law distribution of node degrees – This characteristic directly ties on the previous point. Only very few nodes possess a high node degree, i.e. have links to many other peers. These peers are called hubs or bootstrapping peers. Most peers in Small World networks, however, have a very small node degree, i.e. maintain links only to a few others.

- High clustering – In Small World networks one often finds that two nodes, which have each a link to a third node, are often connected, too. Drawing on social networks, this means that two friends of a person often are also friends among themselves. This characteristic leads to the emergence of node groups which are called clusters.

- Remarkable elasticity and robustness – Small World Networks are usually robust against failures of nodes. That means the overall structure and operation of the system is not influenced if single nodes break down. This quality is mainly driven by the clustering.

Observed characteristics and studied particularities of Small World networks can be transferred to peer-to-peer systems and become interpreted within this new context. Section 3.2 exemplifies which positive effects this phenomenon can cause.

It is important to emphasize that self-organizing processes highly depend on the autonomy of the system's individual parts. With regard to peer-to-peer systems, peers represent these parts. However, autonomy, to the degree required for self-driven structuring of the system, is currently mainly limited to unstructured implementations [Abe02]. Structured systems usually prevent any self-organization since they obey well-defined rules when integrating peers and data into the system. Their overlays usually use DHTs to reference contents. By assigning keys and distributing them to responsible peers, each peer becomes dependent on other peers and looses its autonomy to a certain degree. In the same way, the joining and leaving of peers always involves other peers. If a new peer joins the system, other peers have to adjust their routing tables and adapt their key distribution, etc. Structured systems hence follow a top-down approach by applying their global rules. Unstructured systems are bottom-up driven and only have rules for local peer interactions. Usually, scalability, robustness, and elasticity can be achieved more easily using unstructured systems. Due to their rather deterministic character, structured systems must afford more 'intelligence' to gain similar qualities. Freenet [Dab01] and P-Grid may be seen as representatives of such intelligent approaches. Basically, they try to combine the strengths of both worlds. Although they are structured systems, they make use of self-organizing processes to construct their overlays. P-Grid is introduced in greater detail in section 3.3.

2.4 Summary

Although peer-to-peer has become almost a buzzword in recent times, the concept is not at all a new idea. It already came up with the development of the Internet and corresponds perfectly to the latter's nature of being a distributed information system. However, the following decades and especially the emerging commercialization of the Internet in the 1990s imposed a client-server structure which is just about to recede slowly. In contrast to client-server systems which distinguish carefully between static roles of service providers and service takers, peer-to-peer systems only know equal members that switch dynamically between both roles. The equality of peers makes it principally possible to launch any task on any peer. Thus, peer-to-peer systems are better prepared to compensate failures and disturbances, and are much more robust than client-server applications. Since they don't rely on any central entities for their infrastructure, there are no hot-spots exposed to high workload or network traffic. Hence, single-point-of-failures are avoided and scalability is enhanced.

Peer-to-peer Systems perfectly fit when it comes to connecting distributed content and resources that are valuable in aggregate, but must remain in the custody of their owners. Such constellations are often found in business-to-business applications. There, peer-to-peer technology can give important impulses for new ways of collaboration and information sharing. However, actually in business environments, pure peer-to-peer applications are still relatively rare, and their utility in a corporate environment is still suspect. Doubts for this reservation most likely stem from weak control mechanisms in some implementations that might lead to information misuse and security issues. Hence, peer-to-peer applications targeting the corporate market usually pursue a hybrid approach with central elements to authenticate users before granting access to the peer community. In this domain, hybrid constellations are sometimes called *brokered peer-to-peer*.

Besides connecting and managing large groups of participants, one of the core challenges in peer-to-peer system is information lookup and retrieval. Over the recent years, two major concepts have been developed that tackle this problem. They basically differ in how peers and data are integrated into the system. On the one hand, there are unstructured systems like Gnutella which possess a simple infrastructure and search strategy, but suffer from high network load and incomplete search results. If these drawbacks can't be accepted, structured systems represent the alternative. They use much more effort to orchestratepeers and data to a comprehensive system. Even though this causes higher administrative costs for each individual peer, better and efficient searching becomes possible.

Since, from the view point of this work, peer-to-peer technology is employed to manage descriptions of application services, as they are found in mobile agent systems, the aspect of information lookup and retrieval will represent the focal point of attention within the following chapters.

For the near future, peer-to-peer technology has good chances to broaden its spread. Many name companies have realized the potential of this concept for the contemporary challenges and have positioned their products accordingly. One of the most active players in this group is SUN Inc. Their open source product JXTA [Sei03] represents a unifying approach to promote common standards for peer-to-peer computing in general. In future, JXTA will be able to connect computing platforms of all types and sizes and make them interact as peers. According to Leon Guzenda[2], JXTA represents the turning point for peer-to-peer's breakthrough in enterprise environments. Currently, the JXTA framework already offers many infrastructural services and even addresses such important issues as security and trust management. Future versions will extend these features, and with SUN's partners in industry, JXTA is on its best way find access to many applications.

---

[2]CTO Objectivity, Mountain View CA, USA

3 Selected Peer-to-Peer systems

This chapter gives an overview on three selected peer-to-peer systems and evaluates which one fits best for the utilization in large-scale mobile agent systems to facilitate handling of service descriptions. Limiting the selection to three implementations, may seem to be too narrow at a first glance. However, against the background of targeted environment, this selection explains as follows:

Mobile agents are dynamic software components, employing various application services within a distributed computer network, to accomplish their user-given task. Agencies, the runtime environments for mobile agents, grant access to these services which are usually located on lower layers of the software stack. In other words, agencies provide dedicated communication channels and gateways agents can use to connect to these services. However, the quantity and distribution of services is anything but static. Steadily, new services appear and existing ones are modified or shut down. These dynamics become even more amplified when agencies run on computational devices which are not permanently connected to the network. Laptops, PDAs, and cell phones which may connect only occasionally and even to completely different subnetworks. A fixed binding between user-application features and services on particular agencies seems to be simply unfeasible because of the expected largeness and dynamics of the agent system. Therefore, mobile agents must be capable to adapt to the changing environment. This means in particular, application services must be discoverable even in very large and dynamic systems.

In chapter 2, peer-to-peer systems have been distinguished into three classes according to their basic peer infrastructure. Special roles of peers and uniquely available services, as they are found in super-peer-based and hybrid approaches, always increase the risk of loosing functionality or bringing the whole system to its knees if central components fail. However, due to the depicted dynamics within an agent system, it is impossible to grant permanent connection between agencies and the agent system. Also super-peers, i.e. super-agencies, require a least certain stability of the system to establish backbone connections. Since these demands contradict the properties of mobile agent systems, hybrid and super-peer-based peer-to-peer implementations have not been considered for the selection in hereinafter. Hence, only fully distributed systems have been regarded further.

But this remaining set of implementations shrinks further as soon as the overlay structure comes into play. Looking at the overlay is important, since it crucially determines information lookup and retrieval in the system. Besides the basic peer constellation, now the arrangement of peers and published information decides on an implementation's applicability for a mobile agent system. At one extreme, there are Gnutella-like systems which have an unstructured overlay. They integrate peers and documents randomly into the system's structure, and become better socialized in the system the longer they are connected to it. Searching in such systems is realized by broadcasting query messages to all reachable hosts. This strategy obviously requires a broad network connection, since already the message volume of search requests causes enormous network traffic. Furthermore, undirected searching tends to find preferably very popular contents in the system. Simply because of their broader distribution, it is much more likely to find them than it is for rare contents. Finally, the fact that a peer's quality of integration depends on its time with the system can also be seen as a drawback. All three aspects together thus clearly conflict with the environment mobile agent technology aims at. One can neither assume broad and stable network connections nor long-lasting stays of peers/agencies. Also, even rare application services should be discoverable with equal probability to popular ones. Due to these facts, unstructured peer-to-peer systems disqualify for this book's objective. Only systems with at least semi-structured overlays have been considered for the selection presented hereinafter.

This work tries to prove that peer-to-peer technology can be promisingly utilized for the handling of service descriptions in mobile agent systems. However, since this book only wants to provide a first proof-of-concept rather than a fully-fledged product, only non-commercial and freely available systems have been taken into consideration. Despite all these restrictions, the amount of suitable peer-to-peer systems is still far bigger than this chapter could capture. Therefor, the following three implementations must be seen as representatives of the remaining group of implementations. They have been chosen because they show particularly clear characterizing aspects of several other systems.

3.1 TLS

TLS stands for *Tree-based DHT Lookup Service* and represents a peer-to-peer system developed at the University Reggio Calabria, Italy [Buc04]. This implementation follows the well-known approach of almost all structured systems and distributes a hash table among the peers of the system. As usually, unique keys are stored in the DHT which identify and refer to the available items in the system. The individual parts of the DHT represent the system's global index. Derived keys of search requests can then be utilized to route the queries exactly to the peer that maintains the part of the index which holds the references to the requested content. Depending on its position in the virtual tree, a peer is loaded with a certain amount of routing traffic. Here comes the first novelty of TLS into play. A peer's position within the tree is not fix. The longer it stays connected to the system, the higher it climbs towards the root of the tree, and, thus, the more routing traffic is biased towards it. This approach goes well along with the widely agreed-to assumption that longer connection times give hint to reliable network connections and a certain stability of peers. It is obvious that such peers should be favored to take over more responsibility for the system's overall functioning. However, it also is obvious that the root and the nodes on the first levels of the tree are exposed to a disproportionate amount of routing traffic and are likely to become overloaded. Here it is where TLS' second feature hooks on. To be precisely, peers in TLS are not organized in a single virtual tree, but in a forest of trees.

3.1.1 The overlay structure

Even though TLS actually relies on a forest-like overlay structure, its general functionality can be explained through a common binary tree. In TLS, this tree is called LBT which stands for *Lookup Binary Tree*. Each node of the tree is identified by a unique binary string and represents exactly one peer of the peer-to-peer system. The binary string is determined by the peer's path in the tree.

Beginning at the root, which is identified by a single 1, the path is extended by one bit for each level downwards and depending on the subtrees pursued. That means, a node N on level $x-1$ of the LBT is identified by the string $ID(N) = (1, b_2, \ldots, b_x)$ of length x which represents the path $C = (root, N_2, \ldots, N_x)$. Thereby, $b_i = 0$ if N_i is the left child of node $N_i - 1$, otherwise $b_i = 1$. The depth of a node N in the LBT is calculated by $depth(N) = 1 + max\{depth(leftchild(N)), depth(rightchild(N))\}$, where leaves have depth zero. Fig. 3.1 shows a simple LBT. Each peer's path in given in and its depth next to the rectangle. Information lookup in TLS is ac-

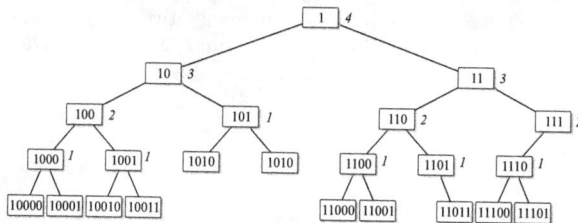

Figure 3.1: A simple Lookup Binary Tree.

complished using a DHT. Each content item that is about to be published in the peer-to-peer system is assigned a binary key first. It is derived using a hash function, here SHA-1. The peer with the lexicographically closest ID to the generated key becomes responsible for it. That means the generated key is sent to that particular peer which then stores the key in its part of the DHT. Every entry in this table represents a key-value pair, linking the generated keys with the IP addresses of the peers which actually store the items. Thus, the item itself and its key are likely to be stored on different peers. To find a particular item in the system, a peer creates a search request. Then, the hash function is applied to the keyword provided which results in a binary string that matches the requested item key. Thus, the search request itself can be routed through the system as it happens with item keys. Beginning at the root, the key determines where the search request is routed to. It will finally reach a leaf node that is responsible for the message's key range. If the requested content has ever ever been published in the peer-to-peer system, the identified leaf stores the references to the appropriate peers. However, here one has to remember that the structure of the tree may change over time what demands for an adjustment of this routing strategy. If nodes join and leave the LBT, levels are added and removed and former leaves may become parent nodes of individual subtrees. Hence, search request will no longer find the key information since it is displaced. The probability that a leaf changes its positions in the LBT generally increases with time. That means, the longer a peer/node is part of the system/LBT the more probable it becomes that other peers are added or disappear what in turn requires the adjustments of the LBT. How routing can be accomplished even within this changing structure is presented in the next section.

If a new peer N wants to join the system, it has to know the IP address of at least one already integrated node A. Starting from the latter's positions in the LBT, N's place in the structure is determined by the shortest subtree of A. That means, starting at A, N will move downwards in the LBT to the leaf node which has the lowest depth of A's subtrees. This strategy avoids the development of deep (sub-)trees and tries to ensure that the entire LBT stays balanced. However, to guarantee global balance, it would be necessary to start the integration process always at the root level rather than at an arbitrarily chosen intermediate peer. Unfortunately, [Buc04] does not address that issue explicitly.

In the case of removing a peer/node of a system/LBT, it will first pass its part of the DHT to its parent node. Thereby it ensures that the stored information does not get lost. Then the node can leave the LBT and its position stays empty or it is filled with another node using standard insertion/deletion operations for binary trees. It is important to remember that the information stored in the part of the DHT is moved-up one level in the LBT. This is where TLS' idea of information aging becomes apparent. Over time, information is shifted continuously towards higher levels of the LBT and thus dislocates further and further from its original position. The higher the system's dynamics, i.e. the higher the rate of peer joining and leaving, the faster the process of information aging. This behavior causes that search requests will find recently published contents better than rather old ones. If the older information should be found instead, the routing scheme has to be adjusted. How this can be done, is shown in the next two sections.

3.1.2 Routing

As outlined in the previous section, the evolutionary aging of data in the LBT obviously causes the problem that search requests are limited to find only the most recently published items. Sometimes, they are even faced with unexpected constellations of peers, e.g. when the targeted node has been removed or displaced in the LBT after a rotation operation. At least the way how information is shifted-up in the LBT ensures that the requested data is always stored at some node on the path from the originally targeted node up towards the root. That means if a search request cannot find the targeted peer at its expected position, the latter's part of the DHT can be found on an intermediate peer on the path up to the root. This path simply has to be followed. If a peer has ever stored the requested key, it will be discovered at some node during the backward traversal of the path.

Routing of search request within the LBT can be accomplished by messages of the following format: $[ID(N_S), ID(N_R), Content]$. By simply using the IDs of the sender N_S and receiver node N_R, a message can be exactly routed within the overlay. The *Content* field stores the actual search request, i.e. the keywords. While routing the message, the three following situations may occur:

The sender's ID is equal to the one of the receiver. In that trivial case, the message has obviously reached its target. Then the peer can extract the keywords of the message's *Content* field, look them up in its local part of the DHT, and reply to the sender with the list of peers (if any) storing the requested content.

The second case is represented by the situation where $ID(N_S)$ is a prefix of $ID(N_R)$. Consequently, N_R must be a node in either one of the subtrees of N_S. To actually locate N_R in the subtrees, only the suffix of $ID(N_R)$ has to be interpreted further. The suffix represents the tail of $ID(N_R)$ where the first $|ID(N_S)|$ bits are omitted. However, no changes are actually made to the receiver's ID. Only the the sender field of the messages is constantly replaced. When starting to route the message, at the first step downwards in the subtree, $ID(N_S)$ is replaced with $ID(leftchild(N_S))$ or $ID(rightchild(N_S))$, depending on the first bit of the suffix which indicates whether the left or right child has to be pursued. Descending that path further, the message's sender field is continuously updated with either one of the child's ID and thereby enlarges by exactly one bit at each level. Finally, the sender's ID will match the one of receiver and case one holds.

The third possibility finally represents the opposite constellation, i.e. $ID(N_S)$ is not a prefix of $ID(N_R)$. In that case, the routing starts by replacing $ID(N_S)$ with $ID(parent(N_S))$. That means the message is routed upwards in the LBT until either case one or two holds. Consequently, the worst case would be that a message is first routed from a leaf up to the root of the LBT and then down again to another leaf. Thus, search costs in LBT can be bound to $O(log(n))$, where n is the number of nodes in the LBT. Since each modification of the sender field is executed by a different peer belonging to the path between the end-to-end sender and receiver, the routing scheme of TLS represents a distributed algorithm.

3.1.3 Adaptability

The outlined routing algorithm reveals that messages are ascending in the LBT if the receiver is not a leaf of the subtree spanned by the sender. Such messages have to be routed upwards in the LBT until they reach a node which is either the receiver or has an ID that is a prefix of the receiver's ID. In the latter case, the found node spans the subtree that comprises both, the sender and the receiver. It is easy to see that messages ascends the more in the LBT the more distant sender and receiver are in the tree. It is obvious that this strategy must lead to a huge amount of routing traffic for nodes that are on one of the top levels of the LBT [Buc04].

To avoid this drawback, TLS modifies its overlay by virtually cutting the first levels of the LBT. Depending on the size of the system, bypass links are created at a certain level from top. The resulting overlay structure is called the *LBT-forest*, since it comprises several subtrees which are all linked in a root chain. Fig. 3.2 visualizes the new structure. These direct links between former sibling nodes represent a shortcut layer for routing traffic. Messages that previously were routed up and down again, crossing the new root chain level, are now forwarded directly to the root of a subtree, containing the receiver node. It is important to mention, that on root level of a LBT-forest, messages can be routed directly between two arbitrary roots without any intermediate hops. That means every root must maintain direct links to all other roots. Whereas this truly accelerates routing and results in better

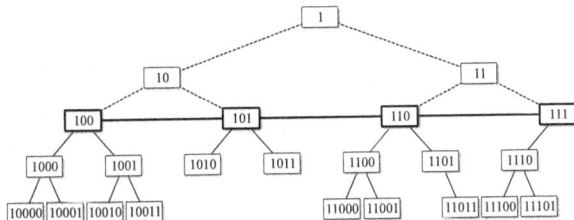

Figure 3.2: A simple example of the LBT-forest.

loadbalacing, one carefully has to look at the parameters which determine at which level the LBT is transformed into the LBT-forest.

It is easy to see that the deeper the level the root chain is established on, the more subtrees the forest structure contains. The more sibling roots exist, the more messages can be routed directly. Driven to the extreme, a root chain on the leaf level of the LBT would allow direct routing with just one hop between any two nodes. However, such a system would quickly break down, since every leaf had to maintain a list to all other leafs. For even medium size systems this can't be a reasonable approach.

Hence, to find the best depth of the introduction of the root chain is always a trade-off which tries to maximize efficiency and speed of routing, and to minimize the risk of the degeneration of the forest structure to a linear list. In [Buc04] it is shown that the level can be derived based solely on the number of nodes in the system. It was presented that for a positive integer constant c and linear growing n the probability that a member of the root chain is involved in routing a message, decreases by $\frac{4}{log(n)}$ if the chain is established on level $l = loglog(n^c)$.

Even with the introduction of the forest structure it is clear that nodes on higher levels still have to handle more traffic. Due to TLS' evolutionary strategy only stable nodes can reach higher levels. This goes well along with the widely agreed-to

41

observation that node stability always indicates a good network connection and sufficient computing power of the node. These capable peers are thus burdened with major parts of the administration and maintenance overhead of the system.

In brief, TLS represents neither a pure nor a super-peer-based peer-to-peer system. The approach has to be placed in between these two categories. Over time, stable and capable peers move to those positions in the LBT-forest which are exposed to higher work loads. In addition, higher peers store information of former peers which have already left the system. These two aspects, influencing the dynamics and structure of the TLS overlay, can be considered as self-organizing and a type of evolutionary adaption of the system.

3.1.4 Simulations

Unfortunately, no implementation of TLS was available when this book was under development. The original paper only presents first results with a prototype not available to public. At least, even at this early stage the main approaches of TLS seem to pay off. Analysis showed that not only the depth of the forest structure could be bound logarithmically, but also the costs for insertion and deletion of nodes. Especially the latter fact highlights TLS from other implementations. Since this makes TLS highly interesting for very dynamic environments, its future development should be pursued.

3.2 INGA

INGA is abbreviated for *Interest based Node Grouping Architecture* and represents a peer-to-peer implementation introduced in [Lös05]. It contains elements of both structured and unstructured peer-to-peer systems, and tries to combine their advantages. INGA does not rely on hash keys which allow direct routing of search request. It rather implements an enhanced unstructured approach. The enhancement has to be seen in the fact that INGA tries to reduce in as much as possible the number of peers search requests are sent to, and thus strives for the routing efficiency of structured systems. Simultaneously, INGA seeks to maximize the autonomy of peers by maintaining only local peer profiles and avoiding administrative overhead. These aspects represent the unstructured influences of the implementation.
The balancing act of combining structured and unstructured features succeeds by using so-called *shortcuts*. Shortcuts are direct links to other peers. These links are grouped in various lists where each list contains peers of a specific type. INGA distinguishes between four types which are introduced in greater detail in the next section. The shortcut approach, together with a corresponding peer classification induces an overlay which has strong similarities to the Small World Phenomenon, mentioned in section 2.3. According to a human peer, each peer in INGA establishes over time a personal expert network with different types of peers, each having

particular strengths. Furthermore, INGA assumes that each peer has a particular profile regarding the content it is interested in. That means, the better and broader a peer's network is, and the clearer its profiles is, the better search requests can routed to promising peers. Peers which have been helpful in the past are believed to be helpful for future requests, too. Their links are continuously maintained in a peer's local lists. Links, however, that did not yield a profit are replaced over time. It is important to state that each peer stores only links to peers which are of its own interest, i.e. have a similar profile and hence similar interests. It is assumed that it is more likely to find desired information on those peers than on others with different profiles.

Consequently, INGA's shortcut overlay causes the emergence of dynamic clusters of peers. Thereby, peers of one cluster have similar interests and are well connected. Therefore, it is more than likely to be able to satisfy search requests within such a cluster. Due to this 'proximity' of peers, response times of INGA systems are quite good.

INGA has been realized on the basis of JXTA and it is focussed on desktop applications that facilitate searching for documents in companies and organizations. It is assumed that members of an organization and employees of a company gather documents about subjects that have a certain relation to their job and position. With this collected information, they will produce new documents that deal with similar topics. Hence, each peer—human as well as software—develops an individual interest profile which becomes recognizable in search requests and a preference for certain contents. Employees, working in similar or related areas and using an INGA-based peer-to-peer client, are able to build-up a cluster and share information very efficiently. An INGA cluster then corresponds to a real-life workgroup.

Even if INGA has been developed for that specific area and, consequently, follows the characteristic constraints of such environments, e.g. strictly local profiles which allow complete control on published content, several of its ideas are unique and can't be found in any other current peer-to-peer implementation. For that reason, INGA's basic approach is briefly introduced in the next sections, and finally evaluated if it also fits for the outlined task in mobile agent systems.

3.2.1 The overlay structure

INGA assumes that information stored in the peer-to-peer system can be classified according to a theme hierarchy. Precisely, INGA uses *Sesame RDF Repositories* [Bro02] and the appropriate query language which. However, it is not important to examine this concept in greater detail here. For more information, please refer to the indicated source.

Such a hierarchy can be compared to the logical structure of a file system which groups thematically related contents and arranges the groups according to their relation. For example, a paper on *Service Discovery in Mobile Agent Systems* could be located in the folder */root/computer/software/agents/mobile* of a local repository. In the same way, any content in INGA is categorized according to the theme hierarchy. To discover items in the peer-to-peer system, search requests must also be formulated according to this hierarchy. That means, if a peer is interested in documents dealing with service discovery in mobile agent system, it creates a search request with the fully qualified category it wants to search in. Such semantic hierarchies are called *ontologies*. They allow in particular the definition of relations like generalization and specialization between any two terms. The longer an item's fully qualified name, the narrower the topic it deals with. In the other way round, cutting off parts from the end of the qualified name, puts the item in a broader context. Hence, one virtually ascends the hierarchy and moves to more and more general levels. For example, if the search request *[Mobile Agent Service Discovery, /root/computer/software/agents/mobile]* for all documents on service discovery in mobile agent system does not discover any items in the peer-to-peer system, the request can be re-posted with the next higher level to search in, e.g. *[Mobile Agents Service Discovery, /root/computer/software/agents]*. Of course, one thereby looses clarity to a certain degree and the request might find less relevant items, but ontologies also represent a powerful means to discover related information one had neither indicated in the search request, nor ever thought of explicitly.

Against the background of INGA's actual target environment, this approach may find documents in a company the requesting employee did not even know about.

As mentioned above, INGA peers are expected to develop a characteristic interest profile which means that they concentrate on individual categories of the hierarchy. This strongly correlates with INGA's application in desktop search. In such environments, users have specific interests and concentrate on some subjects more than on others. Therefore, they have a clear preference for certain contents in the INGA system. An INGA client then tries to discover such peers which match the user's profile or which might be in some other way beneficial for its work. For that reason, the INGA client becomes specialized in a specific area just as its user.

For instance, peers with a preference for the subject Mobile Agents will try to establish shortcuts to peers which either store relevant items directly, or know where to find them. Each peer locally collects this information and stores the links in one of its shortcut lists. To gather new information, each peer filters all communication for addresses of interesting peers. The newly discovered peers are then added to its lists and they are considered for future search requests as well as for message forwarding. Collecting these references is termed *socialization* [Lös05]. Each entry in such a list represents a shortcut—a dedicated, one-hop connection between two peers. INGA distinguishes four types of shortcuts which are all stored in separate lists on each peer:

- Content Provider List – This list contains links to peers which in the past were able to satisfy a search request directly, i.e. stored the content of interest. Future requests on the same or similar topic are most likely to find answer there again.

- Recommender List – If for a particular search request no suitable shortcut can be identified in the Content Provider List, the Recommender List may help. It contains links to peers which requested similar content in the past. The basic idea is that such peers might have already discovered other peers which store the desired information. If so, they will have suitable entires in their Content Provider List. Hence, the local peer should route its request to such a recommender peer which then hopefully forwards the message to one of its content providers.

- Bootstrapping List – If neither a content provider nor a recommender peer seems to be suitable to direct a request to, a good chance to reach a promising peer is to forward the message to a peer which is extremely well socialized, i.e. which has many links to other peers. In other peer-to-peer systems, such peers are known as super-peers or hubs (see 2.1). Links to them are stored in the Bootstrapping List. The idea is that a bootstrapping peer, due to its dense network, might know content providers or at least recommender peers the request should be directed to. The local peer, hence, should route its request to a bootstrapping peer which then can forward the message to a more specialized peer.

- Default List – If the local peer could not find suitable shortcuts in one of the previous lists, the only chance to satisfy the request is to forward the message to a random set of peers. The Default List therefore stores links to peers which are determined by the underlying physical network structure.

Consequently, combining the lists of each peer in the system would result in a four-layer overlay, as it is shown in Fig. 3.3. However, since each peer only has its own lists and therefore a limited view of the system, a local four-layer overlay is induced in which search requests are routed. Searching in INGA, hence, neither follows a key-based structured approach, nor does it implement a Gnutella-like flooding strategy. INGA rather selects a small group of the most promising peers for a specific request, forwards it to them, and hopes that at least one can satisfy it directly, or publishes it within its own expert network. As a result of this strategy, INGA combines low administrative overhead of unstructured systems with efficient techniques for searching like structures implementations possess. The individual shortcut lists of each peer are populated differently. At the beginning, when a peer joins the system, it can use a Gnutella-like strategy by *carefully* flooding the network with search requests according to its interest profile. The responses received come from potential candidates for the Content Provider List. Successfully finished requests can also be used to fill the Recommender List because requests and responses in INGA always contain the complete path they have been routed along. Recommender peers are represented by the second last entry of the request path. This can be easily seen, since the last entry refers to a content provider and thus the previous peer must have been a peer referring to this provider. According to the definition above, such peers are recommenders.

In the same way, also requests of other peers that are routed via any intermediate

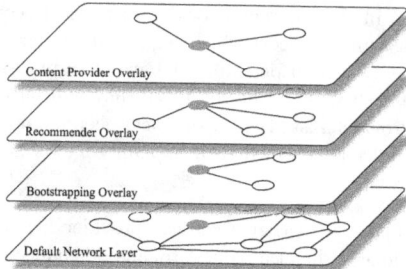

Figure 3.3: The four-layer overlay the INGA approach induces.

peer can be analyzed to extract addresses of content providers and recommenders. The entries of the Default List are implicitly given by the underlying network infrastructure. In particular, INGA uses the set of peers that is provided by the JXTA's rendezvous protocol.

To be finally able to fill the Bootstrapping List it is necessary that every peer in the system monitors its bootstrapping capability. This parameter measures the degree and quality how well a peer is connected to others. Every time a peer starts a search request, it includes its bootstrapping capability in the message. Any peer the message is routed to on its path to the final receiver can use this information to grow its own Bootstrapping List. Please refer to [Lös05] for further details regarding the calculation and capturing of this characteristic property.

3.2.2 Routing

The previous section already touched on the fact that the links in the shortcut lists exhibit different qualities with respect to search requests. Content provider links possess the best quality, since they allow retrieving content with just one hop. Default links hence represent the lowest level of quality, because forwarding requests to such peers is basically not different to the undirected routing in unstructured systems.

Consequently, each peer must be able to identify the best shortcuts for every single search request, regardless if the peer is the initiator itself or if it is just an intermediate node the message is forwarded to. To identify the best links in the local index, an INGA peer implements the following selection scheme:

1. First, the peer checks if the requested content is locally available. If that is the case, it will reply to the caller. Whether or not the content was directly available, the peer checks next if the request has already exceeded its TTL which is measured in terms of hops. If it has, the request is ignored not treated further.

2. If the request's TTL is still below the given threshold, the local peer selects

all shortcuts of the Content Provider List that match the request.

3. If during step 2 too few links were identified, now the Recommender List is examined for links to peers which requested similar content in the past. Similarity is thereby defined according to the underlying ontology.

4. If in the steps before still not enough shortcuts could be gathered which had at least a certain similarity to the current request, step 4 draws on the Bootstrapping List to identify well connected peers.

5. The last chance to select additional shortcuts represents the Default List. Its entries are used to fill-up the missing number of peers the message should be forwarded to.

6. Finally, some of the shortcuts identified during the previous steps are explicitly replaced by some links of the Default List.

The shortcuts obtained after step 6 represent the set of peers the search request is routed to. Upon receiving the message, each one of them will first inspect its own content repository and respond if it stores the requested item. Then each peer will apply the selection scheme above and forward the request if it has not already exceeded its TTL. Step 6 thereby tries to avoid the so-called *Shortcut Problem* [Lös05]. This phenomenon describes a kind of limited sight a peer develops over time if the last step of the selection would be omitted. At a first glance, the peer seems to risk losing some search results by replacing some of the proven links of steps 1-5 with some new and potentially even less valuable ones. In [Mer05], however, it is shown that this strategy is likely to turn out well, since it ensures that new peers with 'fresh' content are discovered. A positive side-effect of this approach is that peers which have just recently joined the system and those which stay connected for only a short time become much faster integrated in the system.

Of course, each shortcut list demands for continuous maintenance. Neglecting this task would quickly result in broken shortcuts pointing to orphaned peers. The peer's index would blow-up, search requests would be sent to peers that have already left the system, and, thus, the network would be loaded with excessive and useless traffic. To circumvent these problems, INGA peers implement an index management which seeks to keep only the best shortcuts in the local lists. Therefore, additional attributes are stored with each shortcut entry. They determine how long the link will be maintained before it is deleted from its list. Among others, the attributes describe when the entry was created, when it has been successfully used the last time, how similar the remote peer's profile is, and how distant it is located. Managing the index basically means interpreting these attributes. With their help it becomes easy to identify old links and those which did not yield enough. Further details on index management based on attributes are presented and analyzed in [Vou01].

3.2.3 Adaptability

In section 2.3 it was said, self-organizing processes always assume a high degree of freedom of the individual parts of the system. In case of peer-to-peer systems this means that peers must be highly independent to emerge self-organizing capabilities. It was shown above that peers in unstructured systems like Gnutella possess the highest degree of autonomy, and hence these systems are able to develop the most evident phenomenons of self-organization.

Although INGA can't be counted as fully unstructured implementation, its peers are almost completely independent. In [Lös05] they are even attributed as *selfish*, since their local index management maintains only those shortcuts which are ultimately interesting to the peer itself. This behavior is compared to a human peer establishing its personal expert network, consisting of people only who are believed to help and give advice in certain situations and on certain topics. Within an INGA system, these connections between humans are mapped to shortcuts between peers. Especially with regard to the envisioned application area—the desktop search in business environments—this correlation is more than welcome. It induces a virtual clustering of employees who work on related topics. INGA clients provide them with the required equipment to find information and documents which are stored on computers of other employees.

The strong similarity to social networks in human societies is unmistakable, and phenomenons of the real life can be observed in the virtual system, too. In particular, INGA networks develop Small World characteristics, as they were presented in section 2.3. Therefore, routing of search requests based on shortcuts is highly efficient, fast, and yields good results.

3.2.4 Simulations

For the assessment of the INGA approach, in [Lös05] a simulation environment is presented which was used for performance analysis in static and dynamic networks, respectively. Particularly, the following values were measured:

- Query recall – This parameter measures the ratio of actually found relevant items and all relevant items stored in the system. A value close to one indicates that almost all possible items were found, whereas a value close to zero represents a poor yield.

- Messages – This value captures the costs of a search request by counting all messages, requests as well as responses, that were necessary to get a certain result.

- Message gain – The ratio of the previous parameters leads to the definition of the message gain. It describes the relation of yield and effort, i.e. how many messages were necessary to gain a certain query recall. High values for message gain indicate that with few messages a high recall was achieved.

- Average path length – Finally, this value quantifies how distant two arbitrarily chosen peers are on average. It is measured in terms of hops.

To be able to compare INGA's results, two other strategies, a Gnutella-like and one following the *Interest-based Locality* [Mer05], were assessed with the same underlying network and identical simulation data. In [Lös05] it is presented in detail that INGA archived the best results in both static and dynamic networks. In fact, the positive influences of the Small World Phenomenon seem to pay-off: INGA networks are characterized by short paths between peers. Hence, a relatively small amount of messages is sufficient to route requests to promising peers. At the same time, the multi-layered overlay proves extremely versatile to identify the right peers for specific routing tasks, which finally leads to good search results. Finally, the high degree of autonomy of the individual peers combined with the self-organizing system structure seems to perfectly fit even to dynamic networks.

Moreover, drawn on the INGA simulations, in [Lös05] several other interesting traits of shortcut-based overlay networks could be derived:

- Bootstrapping peers are able to massively reduce message volume. Due to their high connectivity, bootstrapping peers are able to distribute search requests quickly and efficiently, and simultaneously guarantee a reduction of messages which else were necessary to reach particular remote peers. Their quasi broadcasting functionality must hence be understood as selective distribution capability.

- Extracting information of routed messages expedites the structuring and formation process of the shortcut overlay. Peers should gather addresses of other peers and store these links in local indexes. In the beginning, even links which do not perfectly match the peer's profile should be incorporated. Other peers highly profit of them. Over time, the index management will take care of less valuable entries and strives for slim, efficient shortcut lists.

- Enriching search requests with additional parameters can reduce message volume in the network even more. Since INGA messages store the entire path they were routed along, each peer receiving a message can trace back which peers have been visited already. If its local shortcut index does not contain new links, the message needs not to be forwarded, since that would result in a circular routing which most likely would not reveal any new information.

- A peer's local index works most efficiently if it incorporates in the beginning links to recommender peers which possess a different interest profile. Later on, a peer should concentrate on populating its content provider layer and thereby strengthen its own profile. On the one hand, this strategy truly ensures that a peer develops its own expert network. On the other hand, storing a certain fraction of less profile-related links allows other peers to benefit of this information. The 'selfish' trait of INGA peers is slightly softened here.

3.3 P-Grid

P-Grid, the *Peer-Grid*, is a peer-to-peer system which is developed and maintained at the EPFL [Abe01]. It is a representative of the fully distributed systems and uses a structured overlay. Like many other implementations, P-Grid arranges peers and their content in the overlay according to keys. They, in turn, induce a binary tree structure within the overlay. The mapping between keys and their associated contents is stored in a DHT. The individual parts are distributed among the peer community.

Similar to the implementation presented in the previous section, P-Grid tries to combine the powerful search mechanisms of structured systems with the advantages and strengths of unstructured approaches regarding the basic system layout, i.e. the overlay structure. P-Grid tries to tackle this challenge by replacing any central elements by means of local interaction and randomized processes [Abe03a]. Furthermore, the same source clearly states: '*Also we assume peers to fail frequently and be online with a very low probability.*' Hence, P-Grid seems to be a suitable candidate for dynamic network environments, as they are typical for mobile agent applications.

In contrast to the majority of other implementations, which constrain join and leave operations to one peer each time, P-Grid allows entire subnetworks to freely merge and split [Abe03a]. Additionally, P-Grid sets itself apart by offering range queries to search for content, and it grants efficient operation even for unbalanced search trees or screwed key distributions [Abe03c].

3.3.1 The overlay structure

Each P-Grid peer $p \in P$ is associated with a leaf of a virtual binary tree. Nodes of higher levels are only virtually present. Their roles in the overall tree structure are emulated by the leaf nodes. For more on this, see below. As typical for structured overlays, peers are assigned a unique ID. P-Grid therefore uses binary strings $\pi \in \Pi$. These strings induce the binary tree and hence the layout of the overlay, since they determine a peer's position within this structure. In other words, each peer $p \in P$ is associated with an ID $\pi(p)$, with represents the path from the root of the tree down the leaf. Bit i of $\pi(p)$ thereby codes which subtree at level i has to be followed to reach p; if $i = 1$ the right subtree is taken, else the left. In the same way as peers are assigned unique IDs, each item gets one, too. That means, for each data item $d \in \delta(p)$, the binary key $key(d)$ is calculated, using an order-preserving hash function. Then, the pair $(key(d), \pi(p))$ is sent to a peer r whose ID $\pi(r)$ has the longest common prefix with $key(d)$. This process represents the publication of the item. The peer, this reference was sent to, becomes responsible for it, i.e. it becomes responsible for storing the entry in its local hash table and for making it accessible to search requests. Consequently, each peer stores only such references which are equal or at least prefix-similar to its own ID. Fig. 3.4 shows that peer A with $\pi(A) = 01$ maintains references to all items d whose keys start

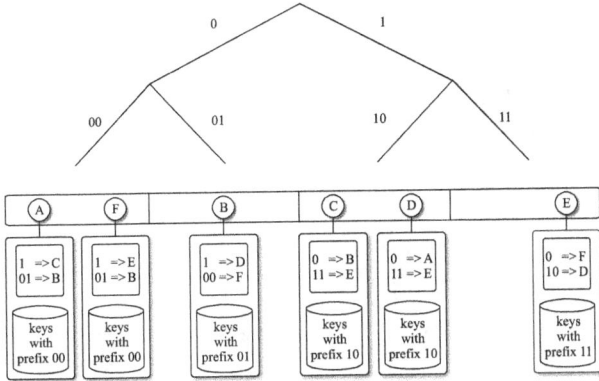

Figure 3.4: Example of the virtual search tree, as it is used to structure P-Grid's overlay.

with 01; peer D, in turn, with ID $\pi(D) = 10$ stores items starting with 10. The same figure shows further that there might be several peers sharing a common key range. Such peers store identical information and are called *replicas*. Replication is a powerful means to enhance fault tolerance of the system and which leads to a certain workload distribution among peers.

How fast items can be published in P-Grid consequently depends on the replication factor f which can be varied for the whole system. At factor of $f = 2$ means, for each key range there are at most two peers responsible. The speed of publishing a single item can hence be bound to $O(log(|\Pi| + f)$. The actual replication process is completely kept transparent. The publishing peer simply submits the key-address pair to one randomly chosen peer of the determined key range. The latter then notifies its replicas through a lightweight gossiping protocol (see section 3.3.4). Therefore, P-Grid probabilistically guarantees consistent data among all replicas. This strategy has been proved to work even in unreliable and dynamic environments [Dat03].

All locally maintained hash tables together represent the global DHT which finally enables peers to search for published items in the system. How this is achieved, is described in the next section.

3.3.2 Routing

Routing in P-Grid follows a simple principle: Besides storing the references to items a peer has become responsible for, it maintains a set of links to other peers in the system. Specifically, for each prefix $\pi(p, l)$ of $\pi(p)$ and length l a set $L(p, l)$ is populated with links to peers q for which holds $flip(\pi(p, l)) = \pi(q, l)$, where $flip(\pi(p, l))$ represents $\pi(p, l)$ with the last bit inverted. In other words, these sets of links represent references to peers which do not belong to the peer's subtree at level l. Due to the opposite bit they belong to the opposite subtree with respect to the level regarded. Consequently, a peer has on all levels at least one reference to a peer of the other subtree. And here it becomes clear what emulation of high-level nodes meant above: All nodes above the leaf level do not represent actual peers. Nodes in the P-Grid tree on levels higher than the leaf layer rather represent exactly the targets for the described link sets. They quasi embody the outlined referencing mechanism, but they are nothing more than a conceptual construct. And this is where P-Grid's routing strategy ties on.

Peer A for example, as shown in Fig. 3.4, has the ID $\pi(A) = 00$ and stores in $L(A, 1)$ the reference to peer C, since $flip(\pi(A), 1) = \pi(C, 1)$, i.e. the two IDs differ in the first bit. Peer D and/or E would have been possible, too, since on the first level, all three are responsible for keys starting with 1. For the second bit, peer A references peer B. Here, there is no choice to select another peer, since peer B is the only one responsible for that key range on the second level of the tree.

With these references for each level in the P-Grid tree, a peer can easily route search requests to the appropriate sites. Upon receiving a message, e.g. a search request, a peer simply compares its ID with its own. If they match, the peer itself is responsible for the key range and will examine its local hash table for references to the requested item. If it finds a suitable entry, the references to all peers actually storing the item are returned to the initiator. In P-Grid, references are IP addresses. However, if the peer's ID and the one of the received search request do not match, another peer has to handle it. Then, the peer will scan its own and the request's ID sequentially. The first bit position they differ in indicates the level in the P-Grid tree the message has to be routed to. The peer will then select one of its references in the appropriate set and forward the request to that peer. The latter will then apply this routing scheme again. Consequently, the message is routed constantly closer to its actual target, since the common prefix of message and individual peers increases at least by one bit at each step. Since P-Grid relies on a binary tree, searching can be accomplished with costs $O(log(|\Pi|))$ in terms of messages. The routing algorithm always terminates successfully, if all peers of the P-Grid system are reachable. Fig. 3.5 shows exactly such a prefix-routing example. As already mentioned in the previous section, P-Grid uses an oder-preserving hash function to derive item keys. This preserving characteristic allows the utilization of range queries to discover desired content. In contrast to standard queries, range queries basically represent a means to sequentially scan a certain range of item keys with just one message. P-Grid knows two strategies to handle this query type. The first is to direct the range query at the beginning to the peer that is responsible for the lower bound of the search range. From there, the message is successively forwarded to each peer until the upper search bound is reached. The second possibility is to route the initial request to an arbitrary node of the key range. Then, that peer will determine the level of the tree the request must be forwarded to, so that each node of the key range receives a copy. In other words, the randomly chosen node in the key range regarded takes care of the distribution of the request. In [Man05] this

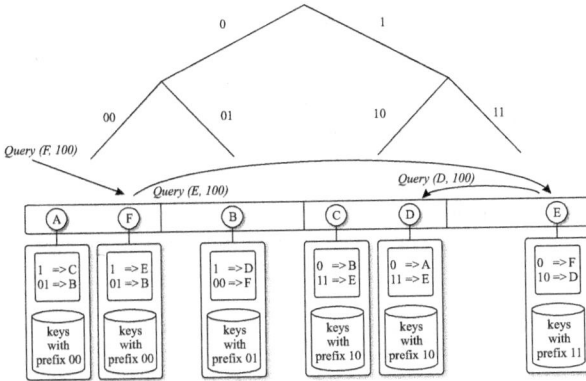

Figure 3.5: Routing a search request in the P-Grid overlay.

feature is presented in detail, and it is proven that costs (in terms of messages) for both strategies are independent of the range size, but solely depend on the volume of the result set.

3.3.3 Adaptability

P-Grid's overlay is created and maintained by a self-organizing process which continuously tries to adapt the structure of the virtual tree to the given peer population and key distribution [Abe04]. That means in a P-Grid system a peer's path is most likely to change over time to fit current conditions. These dynamic adjustments happen strictly local each time two peers meet by coincidence or while actively elaborating the access structure.

In the beginning, each peer is responsible for the entire search space, i.e. it stores all keys that that are published in the system. Whenever two of these undifferentiated peers meet, they start to build the virtual tree by splitting the key space into two parts. Each peer is taking over responsibility for one of them. Therefore, each peer extends its (empty) path by one bit—the first one with a zero, the second with a one. Additionally, each of them stores a reference in its routing table to the other one. Thereby, they ensure that they are still able to cover the entire key space, even if they control directly only one half. A similar splitting of responsibilities always happens when two peers meet which are responsible for the same key range. They would extend their paths and exchange their stored references accordingly, i.e. become further specialized. Besides these basic constellations, several others occur which demand for appropriate behaviors to structure the tree as efficiently as possible. If, for instance, two peers A and B meet and A's path is a proper prefix of B's path, the latter is already further specialized. In that case, A would add one bit to its path which is complementary to the one B's path shows at that position. Hence, this constellation represents a one-side adaption. According to the specialization of A, the peers exchange their keys, if necessary, and add the entries in their routing tables. There are many other situations in which peers simply exchange only routing information, perform replication operations, or recursively delegate adaption to other parts of the tree. For a complete list of all possible constellations see [Abe04]. Important only is here that all decisions regarding the adaptation of the overlay structure are made between two peers and, hence, are completely local. All path extensions and retractions basically try to keep the binary tree as balanced as possible, to ensure best efficiency for all operations on top of the resulting overlay structure. The mutual exchange of routing information thereby promotes a denser coupling of peers and entire merging of hash tables creates replication peers. It has to be emphasized that P-Grid, in contrast to many other implementations, does not use a global replication factor, but again determines the degree of redundancy locally (where the replication factor f represents the upper bound) [Abe04]. This strategy harmonizes well with P-Grid's fully distributed approach and optimally exploits the existing storage resources of involved peers.

Fostering interconnectivity as well as redundancy tries to ensure an optimal workload distribution, and simultaneously enhances the robustness of the network. With the introduction of replicas, a peer's responsibility is distributed on multiple shoulders, so that in case of failure, alternative peers can fill the gap and avoid lost of data. Additional entries in routing tables further guarantee that always several ways exist to reach a certain part of the tree structure. Node failures here too have no influence on the overall functioning of the system.

However, since there is no central control which coordinates the individual adaptation processes, and peers only have a limited sight on the overall system, skewed key distributions and unfavorable decisions might degenerate a balanced tree into a linear list. Fortunately—at least in theory—it could be shown that even in such degenerated overlays search costs (in terms of messages) are still logarithmically bound as long as it can be guaranteed that routing tables contain sufficiently randomized links to others peers [Abe03c].

Almost all current available peer-to-peer systems neglect how to deal with situations when it becomes necessary to update published content [Abe01]. If a peer updates an item, the latter is henceforth available in the updated version. Peers, however, which replicate this item don't become aware of this change. In most systems, a peer has no idea if and how many replicas it has. And hence, there is no way, how it could inform those peers upon a change. Thus, several versions of the same content reside in the system and it is left to higher application layers to identify the desired one. On the peer-to-peer system's level there is no means to assist in this process.

P-Grid, however, uses an explicit update mechanism to circumvent such problems [Dat03]. Decentralization and randomization, as they are utilized for the overlay construction algorithm, thereby play an important role, too. In P-Grid, each peer knows its replicas. Their references are stored in a special list. If an item has been updated, the peer starts its notification algorithm. It is based on a lightweight gossiping approach [Kar00], for which has been analytically proven that it is able to ensure consistent content among replicas. It handles this process with very few messages, but grants fast distribution of the updated information. Hence, like the other features in P-Grid, updating works completely distributed and exploits only local knowledge of peers. It goes well long with the system's core ideas. The existence of the analytical model is regarded to be a significant contribution to research on peer-to-peer systems [Abe01]. Most other implementations must rely on solely simulations results to explain certain behaviors and to assess performance. The same source explicitly states that this update mechanism fits extremely well into highly unreliable and replicated environments as they are found in dynamic peer-to-peer systems.

The last aspect which shall be addressed in this course and which emphasizes once again P-Grid's thoroughly distributed approach, is the challenge of peer identification in dynamic networks. For peer-to-peer systems it is essential that peers can establish connections to others in an ad-hoc like fashion. Therefore, it is necessary that peers have a unique ID in the system. However, since it is most unlikely that a peer is always assigned the same IP in a (physical) network, this property cannot be taken for identification purposes. Of course there are concepts like *Mobile IP* [Per97] and *IPv6* [iiH05] which address mobility aspects in computer networks, but they require changes in the Internet's basic infrastructure and are not accepted broadly yet. Until they find wide audience—if ever—in the meantime, other solutions must be found.
P-Grid tries to tackle this challenge by introducing a '*completely decentralized, self-maintaining, light-weight, and sufficiently secure peer identification service*' [Hau05a]. This service maps unique peer identifiers to dynamically changing peer IP addresses. The basic idea is that peers store the mapping between their IP and ID directly in P-Grid. If they leave the system and get online again, they retrieve this information and update it, if necessary. Although this idea might look paradoxical at a first glance, since actually the system, depending on the mappings, is used itself to store them, in [Hau05a] it shown that this approach succeeds. Under real-world change rates of IP addresses and the online availability of peers, it has been proven, that most of the update processes for the stored IP-ID mappings are completed successfully. Furthermore, those which fail at first and trigger recursive calls, have even a partial self-healing effect on the system structure and make future update processes easier.

3.3.4 Simulations

In contrast to the two previously presented systems, P-Grid can look back on a much longer development process and seems to have gained already a high level of maturity. Consequently, many analysis have been performed which try to capture quantitatively various attributes of the implementation. Thereby, research on P-

Grid often sets itself apart to other projects, by providing mathematical models, backing-up any interpretation, instead of simply drawing on simulation results.

As interesting and convincing many of these evaluations are, as fine-grained and complex are their setups. Generalizing and simplifying particularities would lead to misunderstandings, but to describe and explain all parameters would simply exceed the volume of this book.

Please refer to [Man05], [Mar06], and [Abe03b], where costs for standard as well as for range queries are captured. Also [Abe01] presents in detail which costs searching causes, how reliable it is, and how much overhead must be handled while building-up the overlay. In [Abe04], loadbalancing and replication are of central interest, and [Dat03] presents the above mentioned analytical model for the update mechanism. Furthermore,[Hau05a] assessed P-Grid's ID-IP mapping approach. Finally, there are lots of direct comparisons between P-Grid and other systems. FreeNet [Cla02], for example, which has many things in common which P-Grid and served in parts as reference architecture, is compared to the current implementation in [Abe03a].

3.4 Evaluation

Contrasting all three presented systems, one easily sees that each one of them emphasizes particular aspects. This section hence tries to evaluate which one seems best fitted for application in a mobile agent system to facilitate publication and localization of services on a global basis. In particular, the following parameters are as assessed:

- *Agility* – This parameter evaluates an implementation's capability to rapidly incorporate new peers in and remove old ones from the system's infrastructure. Since peers here represent agencies and are assumed to be highly dynamic regarding their system connectivity, *Agility* should be high.

- *Publication* – Against the background that agencies steadily join and leave the agent system, it is essential how quickly their application services can be published and hence become discoverable. *Publication* evaluates how fast and easy the peer-to-peer system publishes service descriptions in the whole agent system.

- *Update* – This parameter assesses how mature the peer-to-peer system's update features are. If any are implemented, the ease and speed of an update process are evaluated. Since application services of the agent system are likely to change over time, e.g. an additional access point for a certain service might be established on an agency, this parameter has its legitimacy, too.

- *Delete* – In the same way as *Publication* concentrates on how fast information can be published within the agent system, *Delete* evaluates, how fast and how proper such information can be deleted again. If one takes, for instance, an agency which is going to leave the system, it is of high interest to be able to delete all information on its published services to avoid orphaned entries in the virtual maps of the agent system.

- *Search speed* – One of the most interesting parameters is truly how fast service descriptions can be discovered.

- *Search gain* – In addition to how fast a search request will deliver replies, *Search Gain* tries to assess the ratio of the potentially possible and the actually found results.

- *Search efficiency* – Finally, *Search Efficiency* judges which effort has to be afforded to reach a certain quality of search, i.e. how many messages must be employed to yield a high *Search Gain*.

- *Fuzzy search* – It seems more than likely that, against the expected volume and diversity of application services in a global mobile agent system, agents cannot know the exact name of each service type. Moreover, they might have only a vague understanding of what they are actually looking for. Hence, the underlying publication layer (i.e. the peer-to-peer system) should provide something like a wildcard search mechanism. *Fuzzy Search* indicates how mature the system's supports for such things is.

- *Overhead Avoidance* – This value represents the peer-to-peer system's capability to avoid overhead. As sketched in the first chapter, mobile agent applications can best convince in environments which are characterized by unreliable network connections and wireless, slim devices. Therefore, the agent system—and hence the to-be-incorporated peer-to-peer layer—should try to keep network load as low as possible. A high value for *Overhead Avoidance* is desirable.

- *Adaptability* – Because current network infrastructure is changing more and more into a highly dynamic ad-hoc environment, *Adaptability* assess how well each of the peer-to-peer systems is prepared for such application areas. Properties that influenced this value were: How robust is the system against node failures? How mature are processes of self-organization and self-healing? What requirements must be met regarding the stability of the overlay structure and its dependence on peer availability? Are there any advanced functionalities implemented, e.g. caching or replication?

Each of these parameters was ranked on an ordinal scale. Ordinal scales allow basic comparisons of the measured entities. However, they do not allow interval-based operations. In other words, to draw relations like *greater, smaller, more, less, stronger, weaker* is meaningful, whereas conventional addition and subtraction of values are meaningless. That means it makes sense to say: *P-Grid's capabilities regarding Search Speed are greater than INGA's*. It is also allowed to rank candidates with respect to individual parameters as well as to their overall performance, like: *P-Grid's support for Fuzzy Search is stronger than TLS's, but INGA's is even stronger.* However, no distances can be determined between candidates or values, i.e. it makes no sense to say: *TLS is as twice as fast in searching than INGA.*; and neither does: *P-Grid causes half of the administrative overhead than TLS does.* In particular, one cannot add or subtract ordinal values, since it is not obligating for such scales to possess equal intervals between their values.

Nevertheless, an ordinal scale seems to be the most meaningful type of measurement here, since several parameters can neither be assessed numerical, nor can they be captured in the same way for all three candidates. *Adaptability* is one of these rather difficult parameters. Not only that it incorporates several sub-properties, individual setups for each candidate would be necessary to measure them. Just to name one, the notion of replication is truly something completely different in INGA and P-Grid, respectively.

Moreover, some parameters have never been measured so far. In such cases, the author's assessment is solely based on the textual descriptions of the sources referenced within this book. Therefore, the following diagrams use exactly such ordinal scales to *qualitatively* assess the performance of each candidate, and allow relative comparisons between them.

3.4.1 TLS - Focussing speed

The two characterizing properties of TLS are its utilization of keys to identify entities in the system, and its forest-like overlay structure. The former is typical for almost all structured peer-to-peer systems and, hence, TLS inherits many of the resulting advantages. In particular this means that publishing and discovering items—to be more precise, their keys—is highly efficient. Since search requests also rely on keys, they can be routed quickly to the peers which store the corresponding references. If it happens that a responsible peer does not store a desired reference, the request can be routed along the path, ranging from that peer to the root of the virtual search tree. If the desired reference has ever been published, TLS' mechanism of information aging might have shifted it to higher levels of the tree to indicate its probably obsolete character. Nevertheless, if the reference has been published, the request will find it. Even in worst-case scenarios, where a request is routed from the bottom of the tree to its top and back again, search costs are limited to $O(log(n))$, where n is the number of peers in the system. Consequently, *Search Gain* and *Search Efficiency* have been rated high, as shown in Fig. 3.6. Updating, however, is not natively supported by TLS. If a reference has to be modified, the only way to overcome this lack of functionality is to delete the old reference and publish the new one. Deleting, in turn, is basically analogous to searching. Instead of a publishing, a deleting request is submitted to the responsible peer. A major weak point of TLS is that it does not support fuzzy searching. There is even no way to emulate such feature.

The forest structure of the overlay does not only affect the system's *Search Speed* and *Agility*, also it represents a suitable means to promote stable peers to high-traffic positions of the infrastructure. There, they handle large portions of the message volume and establish a kind of backbone communication channel for the system. However, this strategy has to be examined carefully. On the other side of the coin it causes additional overhead in the system. The more dynamic the peers are, that means the more often they join and leave the system, the more frequently stored keys must be moved from lower to higher levels of the tree, since the forest structure constantly changes. Consequently, the more recently an item has been published, the more probable it becomes to find its key at the corresponding node in the forest. Rather old information dislocates more and more from its original site. That means keys, pointing to rare or even uniquely published items, are settled far away from their originally responsible peers. Very popular references steadily find they way back to the leaf level when peers join the system. Driven to the extreme, this strategy might even lead to an accumulation of keys on high levels, whereas the leaves store only very few, very new references. Hence, it becomes more likely that search requests must be re-routed up the tree to actually find the (older) information. Overall search performance hence depends massively on the system's dynamics. It seems inevitable for TLS in such environments to avoid a certain loss of quality in all parameters capturing search performance.

Discovering preferably new content makes truly sense for file-sharing scenarios and similar fields of application. For agent systems though, TLS' aging strategy might have severe drawbacks. Since the peer-to-peer layer here does handle files but service descriptions, the situations changes completely. Generally, mobile agents do not distinguish between old and new services. With regard to published service information in an agent system, this attribute has much less importance than in file-sharing domains. For an agent it is not important if a service was published recently or already some time ago. Even 'old' services possess the same value from the agent's point of view. This problem becomes even more apparent if one thinks of that agencies generally publish their services only when they initially join the agent system.

However, if only few stable but many volatile agencies populate the system, through TLS published service descriptions (in fact their keys) steadily move away from their original position in the tree. New keys, which belong to the same type of service (and hence have the same name), but are provided by agencies which have just recently joined the system, are found easier and more often. Hence, new services repress old ones in the way that they entrap search requests before they can reach old information. This finally leads to the paradoxical constellation where high stability—and therewith high quality—of services result in low probability to discover them. Stable and preferable agencies are quasi forced to re-publish their services according to the frequency new nodes join the system. This would cause additional administrative overhead and network load which has absolutely nothing to do with the agent system itself, but arises solely out of the peer-to-peer layer. Unfortunately, no simulations results are available which could be used to approve or disapprove this hypothesis. Generally, though, maintenance of the TLS overlay is accomplished with low overhead volume. There are neither caching nor replication mechanism in place. Just the messages for the key management are inevitable. Everything on top of this volume is determined by the dynamics of the system, i.e. the frequency of peers joining and leaving the system affects the rate of structural changes in the overlay and adjustments of the key distribution. In the balance of

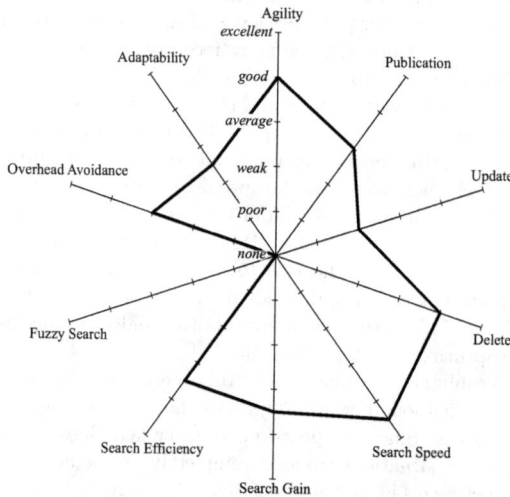

Figure 3.6: The TLS characteristics profile.

things, TLS' expertise has to be seen in its high Search Speed as well as in its simple and clear overlay structure. Due to the latter, overhead costs can generally kept low what fits well to the targeted environment for agent systems. Unfavorably though might become the idea of moving information, since it cannot be simply transferred from file-sharing programs to the requirements of an agent system. Also, the lack of caching and replication might become a weak point.

Based on the current available information on TLS, it seems only be partially suited for applications in mobile agent systems. However, further analysis are necessary for a final decision. Unfortunately, there is still no implementation of TLS available that could be used for additional research besides the contributions of the development team. The efforts on TLS should be regarded further in future. Especially, the forest approach to structure the overly is unique in contemporary peer-to-peer systems and seems to be highly promising to speed-up many standard operations.

3.4.2 INGA - Focussing semantics

The strength of the INGA approach clearly lies in the combination of elements as they are found in structured and unstructured peer-to-peer systems. INGA is able to minimize costs for publishing items and uses completely local profiles, as it is typical for unstructured implementations. Simultaneously, it provides an extremely elaborated overlay and assumes a system-wide ontology which together enables the utilization of search algorithms that are almost as efficient as those of structured implementations. In particular, INGA's overlay allows not only to search for similar content, but even to discover related information which was not explicitly requested in a search request.

Furthermore, the emerging Small World qualities of the overlay lead to short logical paths between peers that have similar profiles and interests. However, even if search speed is convincing, INGA can never reach search quality and efficiency of structured implementations. Its inner system model and the lack of key-based indexing cannot compete with fully structured implementations. Even if a peer's local knowledge base is very broad, temporarily available contents and information of peers which just recently joined the system is hard to capture. This is mainly because INGA uses a kind of rumor spreading mechanism to publish content and interconnect peers. Items can be published and retrieved easier the more messages are exchanged in the system.

Whereas this might be a feasible strategy for INGA's conceptualized domain of application, for mobile agent systems it can't be applied without severe drawbacks, especially in terms of time and network load. Hence, INGA's capabilities regarding publication of information have been assessed very low. Furthermore, although INGA peers can actively advertise their content, they cannot rely on other peers that they actually maintain such references, since each one 'egoistically' stores only links which are interesting to itself. Interest here has to be understood in the sense of a peer's profile. That means INGA assumes a clear interest profile of peers which manifests in the patterns of their search requests. Such preferences are believed to change very slowly over time, if ever [Lös05].

These assumptions go well along applications for desktop-search where users have such distinguishable interests. Regarding agent systems, however, one cannot adopt this idea without restrictions, since it seems to be unlikely that individual agencies develop characterizing interest profiles. This can be emphasized by the fact that an agency and the agent application are two separate things. If ever, an interest profile emerges out of the agent application, but has nothing to do with the underlying agency. However, the latter would actually represent an INGA peer that is assumed to have such a profile. This fact becomes even more apparent if several agent applications with different purposes run on one agency. In that situation a profile would become increasingly lees differentiable. However, the more blurry the profile of an INGA peer is, the less elaborated is its local index, and the less efficient the multi-layered overlay can be exploited. Search efficiency, speed, and gain massively shrink, as shown in Fig. 3.7. The separation of logic into agent application and agency is not found in desktop-search clients, where exactly one user with one profile is directly bound to one INGA client. There, the user's profile seamlessly translates into a virtual profile which in turn can be used to classify the peer and arrange it to a certain layer of the overlay.

One carefully has to keep in mind these fix interests. Although it is shown that

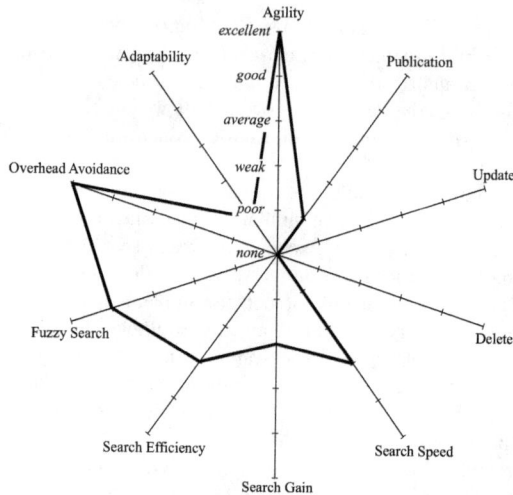

Figure 3.7: The INGA characteristics profile.

INGA performs well even in networks with very low availability of peers and high dynamics, this does not affect a peer's profile which is always assumed to be static. At least the high degree of independence and agility, which makes it easy for INGA peers to join and leave the system, was assessed very high.

All in all, one can summarize that INGA represents a very specialized peer-to-peer implementation, which has strong advantages, but only within narrow limits and restricted to very selected environments. Its strengths mainly depend on two assumptions: A system-wide ontology and clear interest profiles of peers. Neither one of them can be assumed for mobile agent systems though. Especially for large-scale mobile agent systems, establishing a consistent ontology for all services is not feasible. Also, mapping one sub-ontology onto another, or merging them, current approaches cannot satisfy. In the same way, static and distinguishable profiles are alien to agencies.

Consequently, INGA cannot convince in such environments and, thus, the associated parameters show low values. Positively must be admitted that INGA peers fully control their profile which causes very low overhead costs. Message volume, as it arises for remote key administration in structured systems, can be avoided, since no information must be submitted to other peers. This feature might be of interest when it comes to security issues and a peer/an agency may want to control and limit access to its profile.

Even it seems unlikely to induce interest profiles on the level of individual agencies, the INGA approach might provide worthwhile input for structuring higher levels of abstraction. It might be probable that on level of agency groups or domains patterns can be revealed in search requests which could be used as domain profiles. That means, probably an INGA backbone can be employed (additionally) to connect coarse-grained infrastructural elements. Shortcuts would then network such high-level groups which possess similar profiles. Yet, this stays speculation, and since the level of individual agencies is of main interest to this book anyway, INGA cannot convince.

For the future one has to await if multi-layered shortcut overlays can provide valuable input for research on infrastructures for mobile agent systems. Therefore, it would be more than welcome if development efforts around INGA would be revitalized.

3.4.3 P-Grid - Focussing robustness

Similar to INGA, P-Grid seeks to combine the positive aspects of structured and unstructured implementations. In contrast to INGA, however, P-Grid is based on a fully structured overlay which is induced by a classical binary tree and hash keys that determine each entity's position. Thereby, P-Grid inherits from structured systems the efficient mechanisms to publish, delete, and retrieve content. Besides these basics, P-Grid highlights itself by natively providing an efficient update algorithms which even handles the synchronization with replication peers. Consequently, these properties have been rated very high, as shown in Fig. 3.8. Also, P-Grid's means

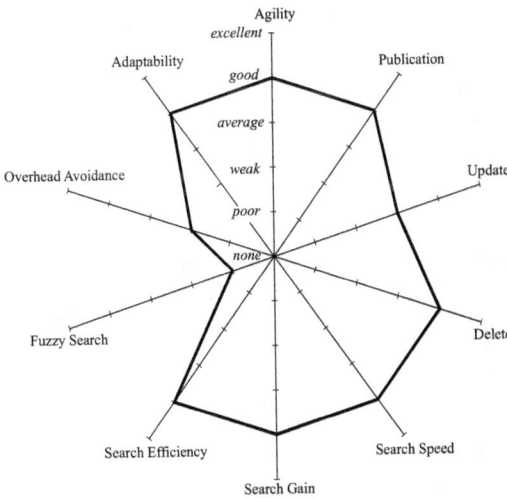

Figure 3.8: The P-Grid characteristics profile.

for fuzzy search are seen as promising. Although the technique of range queries cannot compete with INGA's powerful semantic concepts, they provide a meaningful way to extend regular search capabilities. Additionally, P-Grid's range queries (as well as standard queries) are able to work much more efficient and with higher gain. INGA peers have a deeper understanding about what data is exchanged and can hence abstract from certain notions to find similar and related content. P-Grid peers lack of such a semantic backbone which INGA has with its ontology, but due to the key-based infrastructure, they are more likely to find all contents for a specific request.

Furthermore, P-Grid's strict approach of local interactions and completely distributed control is assessed high. The meaningful combination of randomized algorithms for the construction of the overlay, and efficient search algorithms on top of it, make P-Grid highly interesting for an application in a mobile agent system. The self-organizing and self-healing overlay is extremely robust against node failures and well suited for dynamic, unreliable networks. A nice-to-have feature of P-Grid is its capability of integrating entire subnets into the system structure. This quality led to equal grades in Agility for both implementations, although it has to be assumed that integrating a peer in TLS is slightly faster than in P-Grid because of the latter's replication mechanism. Besides these core properties, P-Grid addresses several other aspects which make it further interesting. For example its gossiping algorithms for handling synchronization between replicas operates even in very unreliable environments as they are found in ad-hoc networks—a typical application area for mobile agent systems. Also, the ID-IP mapping mechanism is best suited for areas which are characterized by very low peer availability. Of course, all these advantages are not for free. Especially the replication mechanism and the random-driven construction of the overlay demand for additional message overhead to compensate the limited sight of the individual peers. Hence, this property has not been rated top.

In brief one can conclude that P-Grid represents a convincing compromise of features, as they are found in structured and unstructured peer-to-peer systems. The key-based search guarantees that contents can be published fast and discovered efficiently. Even rare items are not suppressed. Updating and deleting mechanisms are provided, and self-organizing processes constantly adapt the overlay structure to the key distribution and peer availability. In comparison with the two previously presented implementations, P-Grid is truly the most mature one. Almost every aspect of the approach has been addressed in literature and most of them have been studied analytically. Moreover, P-Grid represents a vital project, the system is constantly maintained, and the source code can be obtained for research activities. In future one has to expect further significant extensions to the system. Among others, trust and reputation management as well as semantic searching are current topics for upcoming versions.

As any current available peer-to-peer system, P-Grid too emphasizes selected aspects more than others. An entirely well-suited implementation for an application within a mobile agent system is not available so far. Although P-Grid cannot fully convince in all directions, it provides the most promising bundle of functionality for a feasibility study and the implementation of a prototype. The next chapter shows how P-Grid was engaged to network isolated agencies to a coherent agent system, and how publication of service descriptions can be realized using this infrastructure.

4 APLICOOVER

This chapter introduces APLICOOVER, the software component which has been developed in the course of this work. APLICOOVER stands for *Agent Platform Interconnecting Overlay* and represents a P-Grid-based peer-to-peer system that can be utilized to network individual agencies to a comprehensive agent system. Additional to the pure linkage, APLICOOVER is able to publish and discover service descriptions within the community of agencies. Hence, it represents a suitable means to retrieve any kind of information on application services which can in turn be used to create the virtual maps of the agent system. Provided with such maps, mobile agents are then able to navigate within the agent system autonomously and without human intervention. APLICOOVER therefore represents the next building block towards agent systems of the second maturity level which allow service-oriented migration.

4.1 A transparent two-level approach

In section 1.2.1 the potential of mobile agent technology has been portrayed drawing on a sample scenario. A crucial role thereby played the notions of a virtual map of the agent system, a route planner service, and a migration optimizer. The map was used to spot application services necessary to the fulfillment of a mobile agent's given task. Based on the identified locations in the agent system, the route planner and migration optimizer were utilized to calculate the best route through the network and the optimal migration strategies, respectively. Erfurth implemented with *ProNav* a framework that comprises exactly theses three entities [Erf04]. It can be employed to form an additional software layer between the agent platform and the agent application of an arbitrary agent system to extend its functionality with the mentioned services. With ProNav, mobile agents can use the map to discover required services in the agent system, and engage the route planner and migration optimizer to navigate to the desired sites. This way of autonomous and service-oriented migration characterizes, according to Erfurth, an agent system of the second maturity level. It differs from systems of the first level mainly in the strict avoidance of human intervention during the selection and navigation process. Agents in systems of the second level utilize software components like *ProNav* to select their desired services and identify suitable paths through the network.

It has been shown that the quality and actuality of map information directly relate to the efficiency of mobile agents. Obsolete information and erroneous records may lead to situations in which agents try to migrate to temporarily unavailable agencies or which have ceased to offer a certain service. To circumvent such problems, map information should always represent the most recent and best possible snapshot of the agent system. However, these claims interfere and contradict. For example, the overhead network traffic would tremendously increase the more accurate and recent the map information is expected to be. Hence, mobile agents themselves would be impeded during migration processes which would lead the idea of optimal routing strategies to the point of absurdity.

Consequently, it becomes inevitable to reduce the data volume for the creation of the maps. Erfurth hence introduces fish-eye-view maps which contain sharp information for each agency's proximate surroundings and blurry information for distant areas of the agent system. Section 1.2.2 showed that such maps are still sufficient to gain the desired level of service-oriented migration effectiveness while simultaneously reducing administrative network traffic. Erfurth used the agent toolkit Tracy to prove its concept. However, Tracy in its current stage of implementation induces a hierarchical infrastructure of the agent system which leads to the emergence of central elements. A failure of these important parts brings the whole agent system down to its knees.

Therefore, recent research efforts at FSU pursue a new direction to bypass these weak points in Tracy's current infrastructure. In principle, it is tried to emphasize the peer character of agencies and to replace any central and hierarchical elements with a peer-to-peer-like orchestration of agencies.

In contrast to the classical understanding of peer-to-peer systems, as it can be found with file-sharing applications, the notion has to be re-interpreted in the shade of mobile agent technology. Other than in file-sharing programs, where individual files represent the unit of information, here, with the proposed application of peer-to-peer technology in a mobile agent system, the units are service descriptions. Not only that these units do not have a direct counterpart in the filesystem, the way of searching is fundamentally different in both worlds. Within the file-sharing world, single, very specific requests which come one-by-one trigger the application to search for the corresponding files in the system. In other words, searching usually happens sequentially and only as a consequence of a concrete request—that means reactive. Within in agent world, service descriptions demand for other ways of handling. Since the virtual map of the agent system is shared by all agents of a platform, the indexed information must be as manifold as the tasks of the agents are. Additionally, the map service must gather such information already in advance to requests of the agents, since a reactive mechanism would cause delays. Hence, searching for service descriptions must happen proactively and prior to concrete requests of agents.

Since it cannot be foreseen if and when a particular service will be of interest to an agent, all available descriptions should be gathered and indexed in the map.

One easily sees that in context of a mobile agent system the claims to peer-to-peer technology are completely different; especially regarding time bounds. Whereas in file-sharing applications time limits usually do not play a crucial role, they definitively do for mobile agent systems. As mentioned above, obsolete map information may lead to a loss of efficiency in agent activity because of the raising probability to encounter broken network connections as well as downed agencies and services. Hence, a peer-to-peer system, serving as 'glue' between the distributed agencies, is much more time-driven than it would be in other areas this technology is engaged [Doe05]. To fulfill these time constraints besides the pure handling of service descriptions, the same source proposes a two-level approach. Analogous to the two parts of fish-eye-view maps, an agent system is distinguished in a local and a global level of abstraction.

On the local level one finds two components which network agencies to so-called *regions*. Similar to Tracy's current notion of a domain, a region represents a cluster of agencies. They are interconnected through the *QuickLink* component. On top of this basic infrastructure, the second module facilitates the management of service descriptions within such a region. This component is called *ServiceJuggler*. The step from this local level to the global one is finally realized with APLICOOVER. It networks isolated regions to a coherent agent system and allows the management of service descriptions on this global level.

All three components are implemented in the Java 5.0 programming language and can hence be engaged by a variety of agent systems to extend functionality. The three modules are presented in greater detail next.

QuickLink. It is obvious that any dynamics of the agency population can be better detected in local environments rather than in the entire agent system [Doe05]. Hence, QuickLink makes use of this observation and connects agencies of an IP-subnet to a local group which it referred to QuickLink group or region hereinafter. QuickLink manages any joining and leaving of nodes within a region, and structures this group as a logical ring topology. To maintain and administer its region, a QuickLink module uses UDP broadcasts on two ports: the *cycle port* and the *update port*. Since local networks generally have reliable and sufficiently broad connections, and the broadcast messages can't leave the local subnet, this method represents an efficient means to reach all agencies of a region with little effort.

A broadcast on the cycle port basically conveys a list which contains the IP addresses of all group members. Upon receiving this broadcasted list, each node compares it with the last received and updates the latter if necessary. Hence, each agency is always aware of all members of a QuickLink group. A broadcast on the cycle port is initiated periodically, but always by a different agency of the group. The sequence is determined by each agency's IP address. When receiving a new list, an agency simply has to examine the sender's address to know who's next in line. The agency with the successive IP address is responsible for the next broadcast message once the cycle time is over.

Before a new agency may join a QuickLink group, it first has to listen for broadcasts messages in the local subnet. Therefor it listens on the cycle port. If the new agency does not receive a broadcast message during the join time waiting period, it builds-up its own QuickLink group. In case the agency which is about to join the region receives a broadcast during the initial phase, it adds its own IP address at correct position in the list, and submits the updated list to all members of the group via the update port. The already established members of the QuickLink group update their own lists accordingly and thus integrate the new agency into the group. Henceforth, the new one takes part at regular broadcasts as any other agency does.

Usually, like in other network systems, a node will leave a QuickLink group rather unpredictable. Often a node will virtually seem to disappear because it has lost its connection to the network. If the successor of such a silent node does not receive the broadcasted IP list, it will wait another cycle before it initiates its own broadcast. The list that is emitted marks the silent peer as missing. The successor, that means the second successor of the silent node, then tries to reach the latter with a TCP Packet InterNet Groper (PING) to verify its unavailability. If that succeeds, the mark in the list is removed and the cycle continues as if nothing has happened. However, if even the TCP package cannot reach the agency, it is assumed that it has lost connection and its marked entry is removed from the list. This delayed deleting with a 'reprieve' of one cycle takes care of the fact that QuickLink uses UDP packages for list broadcasts. Since this protocol does not support explicit collision detection, as TCP in contrast does, it may happen at times of high network load that some packages get lost during a broadcast. Therefore, QuickLink tries to redeem this uncertainty with a waiting state and a TCP connection attempt. Only if that fails too, the agency is assumed to be out of the region and the corrected IP list is broadcasted to the remaining members.

In principle, a QuickLink group can have an arbitrarily large number of members. However, to be able to react adequately to the dynamic changes within the agency population, a group should not be larger than a hundred nodes [Doe05]. With that size and a cycle period of one second, new agencies can be integrated within one or two seconds, and left members can be detected in less than two minutes. Furthermore, with a group size of roughly a hundred members, broadcasted lists can be kept small enough, so that they don't load the network excessively.

ServiceJuggler. The second component on the local level is ServiceJuggler. It is located on top of the logical network that has been established through the QuickLink module. Other than the latter, ServiceJuggler does not address the basic problem of interconnection, but manages the descriptions of the application services within a QuickLink group. Therefor, it launches a directory service on selected agency of the group. Usually, this is the most stable agency with the highest processing power. To determine that agency in the local region, ServiceJuggler draws on the broadcasted QuickLink lists. Besides the IP addresses of the group members, they contains some additional information on each agency's physical 'fitness'. How these values are calculated it not relevant here. For further details, please see [Spr05]. Important only is that ServiceJuggler is able to spot the most capable agency of the local group. The identified candidate then sets-up the directory service which then is used to store the service descriptions. To make sure that only one directory is maintained within the group, the dedicated agency is marked in the broadcasted QuickLink list. The agency running the directory service is called *region manager* which is basically the same concept as Tracy's domain manager, as it referred to when depicting Erfurth's approach in section 1.2.2. Once the directory service is up and running, all prerequisites are in place and the service providers of the local region, e.g. stationary agents, can register their services in the directory. If it happens that the agency's performance decreases in the course of time, and if it can be even foreseen that it will shortly leave the QuickLink group, ServiceJuggler can start a new directory service on the next suitable agency and transfer the stored information from the old one. Hence, ServiceJuggler in able to adapt to the changing network constellations it is faced with.

Equipped with this intelligent directory service, all agencies of the local group can query the directory for service descriptions to build the maps, as they are required by mobile agents for service-oriented navigation within the agent system. However, so far, only the region is covered and indexed, that means only the local portion of the fish-eye-view maps can be created. The outer border which covers the profiles of distant regions is still a white area. To fill that part too, an agency must be able to retrieve information from remote regions. At this point, the transition to the global level becomes necessary.

APLICOOVER. On the global level finally, APLICOOVER enters the stage. It represents a component, providing the functionality to connect the individual QuickLink regions to a coherent agent system, and to publish service descriptions beyond local boundaries. In other words, whereas on the local level infrastructural interconnection and information management is clearly separated between Quick-Link and ServiceJuggler, APLICOOVER on the global level takes care of both tasks. It encapsules and uses P-Grid to establish a peer-to-peer overlay network of agencies. APLICOOVER is launched on an arbitrary agency of each QuickLink group and gets its information—the region profile—from the region manager of the group, i.e. from the agency running the service directory. Both can run on the same machine, but they need not, since the communication channel between them is based on RMI which allows inter-computer communication.

The region profile is a reduced and condensed snapshot of the available application services of a region. This data reduction is necessary to make the approach scale in large systems. Although APLICOOVER, and thus P-Grid, would be principally able to handle the flood of all service descriptions, efficiency would quickly decrease. The reduction is realized by the region manager by bundling similar service descriptions. If, for instance, a service is available twice in the region, the information to be published through APLICOOVER would contain only that the service is available, but not on which agencies it runs. This reduction of information corresponds to the character of fish-eye-view maps. Agencies can retrieve information from remote regions by searching for service descriptions via APLICOOVER. This information can be used to fill the white borders of the maps. Individual remote agencies thus never appear directly in this area—otherwise that part would not be blurry. The blurriness results of the rather coarse-grained information of the region profile which is published through the region manager. It virtually covers its associated agencies towards the global level. Hence, a particular service of a remote region is indexed in a map only with its name, its cardinality, and the address of the associated region manager. A mobile agent which wants to use that service would only be able to identify the remote region manager in a first step. In a second step it has to migrate to that remote manager and use its local map to finally locate the agency actually hosting the service. The second map contains the blurry point, which has been identified in the first map, in detail, since it represents in both cases the same information. However, whereas in the first map it was located in the blurry boundary, it is in the center of its home region map. Consequently, the mobile agent can draw on the detailed descriptions, spot an agency, and utilize the service of interest. Fig. 4.1 shows the logical infrastructure of the envisioned agent system. QuickLink and ServiceJuggler first integrate all agencies of a local subnet to a region. These regions are then networked to the agent system engaging APLICOOVER. Of course,

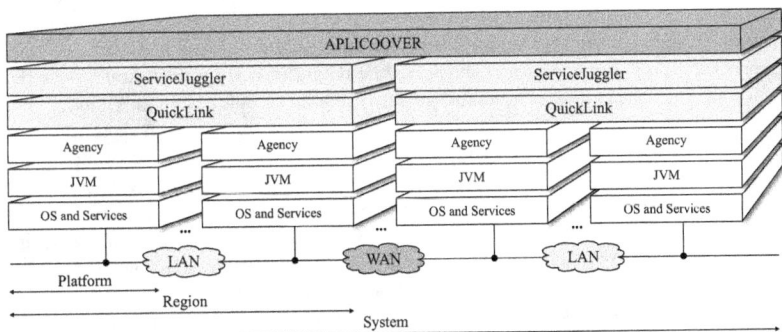

Figure 4.1: A logical view on APLICOOVER and the other infrastructural layers.

also in this two-level approach, particular roles of agencies can be identified. At a first glance, especially the region manager seems to resemble fully the domain manager whose role has been criticised in previous sections. However, this new approach reveals such roles less obviously to the agents, and allows a much higher flexibility in adjusting the infrastructure. In other words, QuickLink as well as ServiceJuggler have the suitable means that guarantee a region's operational functionality even if dedicated agencies (e.g. the one hosting the directory service) fail. Even on the creation process of the maps, the dynamic adjustments of QuickLink and ServiceJuggler have no influence. At every time it is assured that the central portion of the virtual maps can be filled no matter where the directory service runs. On the global level finally, no distinguishable roles can be found at all. All region managers are treated uniformly by APLICOOVER and get interconnected to the fully-fledged agent system. Particular structures and tasks, like the P-Grid infrastructure or replication processes, are completely hidden to the agents. Hence, both levels gain a much higher robustness and represent a promising approach for the next generation of Tracy's infrastructure. Most importantly, QuickLink, Service-Juggler, and APLICOOVER allow the efficient management of service descriptions, as they are needed for the service-oriented migration of mobile agents.

The following sections present the APLICOOVER module in detail and show how it collaborates with the other layers of the software stack.

4.2 Design

Based on the descriptions in the previous section, it is obvious which tasks APLICOOVER addresses within the envisioned infrastructural concept. The most important ones are shown in Fig. 4.2. Dawning on the descriptions, three actors can be identified. First, this is the underlying agency, hosting the APLICOOVER component. It seems reasonable that the agency is responsible for starting and closing the module. Also the adjustment of parameters it should be its tasks. ServiceJuggler is the second actor. It represents the main component from which APLICOOVER receives its data. ServiceJuggler engages APLICOOVER to publish and delete the region profile, respectively. It should also be possible to publish, modify, and delete individual service descriptions of a profile. The third actor can been seen in agents which seek for services in the agent system. This could be a stationary map agent, for instance, which is about to create the fish-eye-view map of the system and queries APLICOOVER to fill the white boundary areas. However, it might also be any other (mobile) agent that utilizes APLICOOVER directly to find particular information. In either case, four types of search methods are provided by the module. On the one hand, these are the standard search and the range search, as they have been introduced in section 3.3 These two methods are directly supported by the encapsulated P-Grid system. On the other hand, it seems to make sense to cache search results for subsequent requests. This would have a positive impact on network traffic and response time of the module. Finally, it should be left to the agents if and when they actually want to utilize gathered results. Hence, a separate use case has been modeled for the retrieval of the information. In particular this means that an agent can trigger a search request which is then processed asynchronously by APLICOOVER. In the meantime, the agent can put its attention on other tasks and does not have to wait for the response form the global layer. Of course it should be able to inquire the status of its search job and any time, and pick up the results as soon as they become available.

Drawn on the general idea how APLICOOVER has to be integrated in the infrastructural concept of the agent system (see Fig. 4.1) and on which interactions arise in collaboration with ServiceJuggler and other components, Fig. 4.3 gives a coarse-grained view how the APLICOOVER module has been designed to fulfill its purposes. The APLICOOVER layer forms the middle tier in this design. It utilizes services which are provided by the P-Grid framework and offers in turn its own interfaces towards the bottom tier where they are employed by ServiceJuggler an

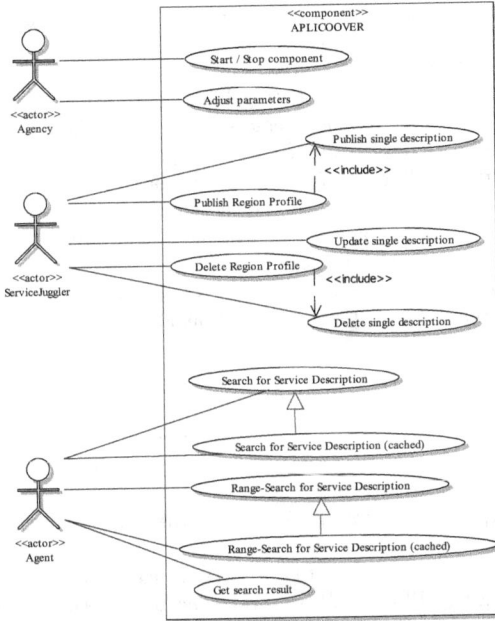

Figure 4.2: Derived use cases for the APLICOOVER component.

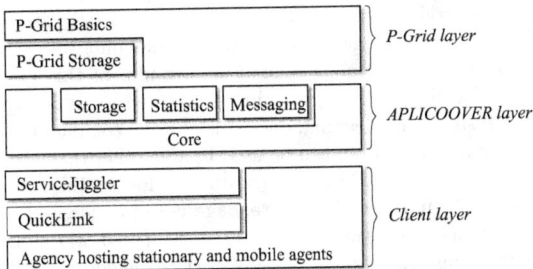

Figure 4.3: A high-level view on APLICOOVER's inner structure.

even lower layers of the software stack. The *Core* component provides these interfaces and covers thereby the inner structure of APLICOOVER as well as P-Grid's. Consequently, it is easy to replace certain submodules or put a different peer-to-peer implementation on top.

Much in the same way as the *Core* component pretends a coherent layer towards lower layers, it distributes the individual tasks between dedicated subcomponents towards higher levels. Thereby, *Storage* is the most important one. It handles the entire data and message traffic that arises with publishing, updating, and deleting of service descriptions. Therefore, *Storage* interacts directly with the P-Gird storage layer. *Messaging* is APLICOOVER's specialized component for any communication beyond the fraction that results from *Storage*'s work. It hence connects to the P-Grid basics layer which is responsible for such tasks. *Statistics* continuously monitors various performance parameters—APLICOOVER's as well as P-Grid's. It communicates with *Storage* as well as *Core* and sticks to the P-Grid basics layer, too. The *Core* component embraces all these submodules and takes care of their proper registration within P-Grid. It employs functionality of both the P-Grid storage and basics layer, respectively.

Fig. 4.4 shows the Java classes that have been derived from this design and some exemplified connections to the P-Grid layer. The graphical arrangement of the classes tries to correspond to the presented architectural model. Attributes and method parameters are hidden in favor of clarity. For this information, please refer to APLICOOVER's API documentation. All classes, their functionality, and their role in the overall design are explained subsequently.

RemoteAdministration. For connecting to ServiceJuggler, lower layers of the agency, and mobile agents, APLICOOVER provides the interfaces `RemoteServices` and `RemoteAdministration`. They offer all methods which are visible to the bottom tier. In case of `RemoteAdministration`, these are at first the methods `init` and `shutdown` which can be used to launch an APLICOOVER instance and attach it to an existing agent system, and shut it down again, respectively. These methods are directed to either ServiceJuggler or the underlying agency. Clearly to ServiceJuggler belong the services `publishRegionProfile` and `deleteRegionProfile`. The former fetches ServiceJuggler's region profile and publishes it in the global agent system. Invoking `deleteRegionProfile`, results in this profile's deletion from the system. Usually, publishing should be triggered when the agency (i.e. the region) joins the system, and deleting, once it leaves the community. The methods `getCacheTime` and `setCacheTime` can be used by the either the agency or ServiceJuggler to adjust APLICOOVER's caching features. More on this is described in the following sections. The purpose of `routeMessage` is to provide a universal channel for message communication between agencies via the global overlay network. However, in the current implementation stage of ServiceJuggler and APLICOOVER, this method simply demonstrates the possibility of such features. Since ServiceJuggler can not handle incoming messages so far, they are either printed to a text console or simply omitted in APLICOOVER. It is left to future versions to reveal

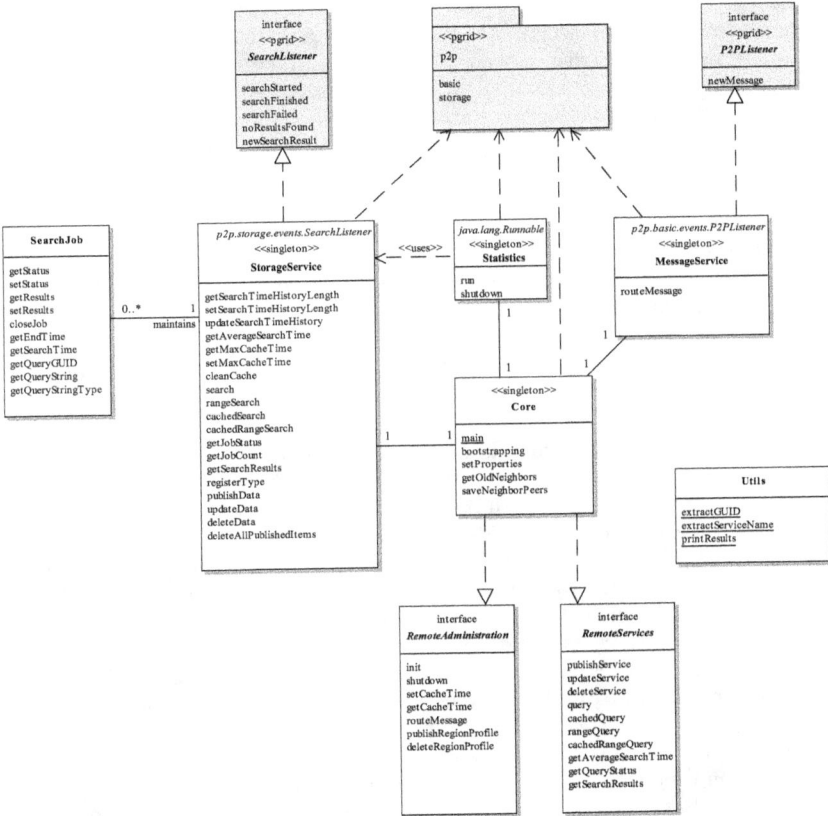

Figure 4.4: The classes of the APLICOOVER module (white) and selected components of the P-Grid framework (grey).

the full potential of this feature and benefit from this advanced functionality.

RemoteServices. This interface represents the second part of services that are provided by APLICOOVER to lower layers. Whereas `RemoteAdministration` concentrates on rather coarse-grained tasks, `RemoteServices` provides fine-grained functions. The most important ones are `publishService`, `updateService`, and `deleteService` which facilitate the handling of single service descriptions, in contrast to the corresponding methods of `RemoteAdministration` which always address a whole bundle of descriptions. The `publishService` method publishes a single description, `updateService` allows the modification of an already published service, and `deleteService` deletes the description in the peer-to-peer layer so that it can't be found any more. The remaining methods of this interface are related to searching: `query` and `rangeQuery` can be used to issue new search jobs, seeking for service descriptions in the agent system. Much in the same way, `cachedQuery` and `cachedRangeQuery` do the same, but also scan the APLICOOVER cache to expedite processing of search jobs. Finally, `getAverageSearchTime`, `getQueryStatus`, and `getSearchResults` are employed to monitor the status of a search job and retrieve its results. All methods that are related to searching are usually triggered by agents. For APLICOOVER it makes no difference if such invocations come directly from mobile agents, or from stationary entities like the map service.

Core. The `Core` class implements all methods which are defined by the two interfaces mentioned above. All of them are exclusively accessible via RMI, i.e. all communication with ServiceJuggler, the agency, and the agents are managed over RMI. The connection to the P-Grid layer, however, is established with regular method calls. The RMI-based communication channel towards lower layers allows it that APLICOOVER can be hosted by different agency than ServiceJuggler's directory service, but both can still interact. In the same way, all agents of a region—mobile as well as stationary—thus gain direct access to APLICOOVER and can use its services. As already mentioned above, the `Core` class is responsible to properly link the APLICOOVER and P-Grid tiers and establish the connection to the peer-to-peer system, i.e. the agent system. Therefore, `Core` implements the methods `init` and `shutdown` of the `RemoteAdministration` interface. The first one uses in turn `setProperties`, `getOldNeighbors`, and `bootstrapping` to initiate exactly this linkage process. If no systems exists the local region can connect to, APLICOOVER creates a new system (with just the local region) and listens for connection attempts from remote APLICOOVER instances. Although, `Core` implements the omitted methods too, they mainly delegate invocations to methods of the specialized classes and basically (un-)wrap parameters. The computational logic of these methods is hence kept in the corresponding methods of the other classes and described below. The static method `main` finally, is executed when APLICOOVER starts-up. It takes care of the module's proper registration in the RMI registry and grants accessibility to ServiceJuggler and all other components. It has to be emphasized again that `main` only instantiates the APLICOOVER object. First with the invocation of `init`, a connection to the global agent system is established. The `shutdown` method cuts this connection once APLICOOVER shuts down and deregisters the component in the RMI registry. Hence, after `shutdown` has been executed, all subsequent communication attempts with APLICOOVER will fail.

MessageService. The `MessageService` class offers only two methods in its current implementation stage. The first one is `newMessage` which results of the implementation of P-Grid's `P2PListner` interface. This interface is part of the P-Grid basic layer and realizes a kind of call-back mechanism. By implementing this interface and registering later at a dedicated access point, `MessageService` receives all messages that do not directly result from `Core`'s and `Storage`'s work. The counterpart to this method is `routeMessage` which receives a delegated invocation from the corresponding method in `Core`. It represents the outbound communication channel which allows sending of messages `newMessage` can receive.

StorageService. As the amount of methods already suggests, the major part of APLICOOVER's functionality is realized with the `StorageService` class. It handles the whole data and traffic volume that arises from with publishing, modifying, and deleting of service descriptions. Before any kind of information can be published via P-Grid, it must be registered with a type. Similar how webservers distinguish types of information like `text/html` or `image/gif`, P-Grid also characterizes content by a type. However, with APLICOOVER this type can be set freely; `text/plain` is as much correct as `description`. This particularity results from P-Grid's major area of motivation: file-sharing applications. There of course, it is highly welcome to distinguish between different file types. For P-Grid's application within an agent system, this feature could be exploited in future versions of ServiceJuggler to classify and categorize service descriptions and, hence, pave the way for enhanced search techniques. Although APLICOOVER's search methods are already able to distinguish different types, ServiceJuggler does not yet provide such information within the region profile. Therefore, currently only one universal type is assumed for all descriptions.

The functionality of `publishData`, `updateData`, and `deleteData` is straightforward. They are used to publish, modify, and delete single service descriptions in the agent system. The method `deleteAllPublishedData` completely clears a region's profile in the agent system once a region decides to separate.

The second part of `StorageService`'s methods addresses various types of searching. The methods `search`, `rangeSearch`, `cachedSearch`, and `cachedRangeSearch` realize this functionality. They are triggered by the different query methods of the `Core` class. The first two of them create a new instance of class `SearchJob` upon receiving a new request. A `SearchJob` object stores all relevant information of the search as well as any results that arrive. All jobs are maintained in `StorageService` using a so-called *cache*. Successfully finished jobs remain in the cache, even if their results have already been fetched and the original initiator of the job is no longer interested in it. This is done because it is assumed that other queries might request similar information. Engaging this local cache, such requests could be processed much faster than searching for information in the global peer-to-peer layer could be done. This is exactly what `cachedSearch` and `cachedRangeSearch` do. If they are used to search for service descriptions, they will first scan the cache for jobs that already gathered such information. If that succeeds, a reference to the cached job is returned to the caller. If no suitable jobs could be found, a flag, passed as a special parameter, determines how the request is processed further. If it is not set, the search ends and a null-value is returned to the caller. However, if it is set, a new `SearchJob` object is created and the search is continued globally.

It was intentionally avoided to engage the cache for all requests. This gives a caller much more flexibility, since it can decide which strategy to take for each single request. The standard methods (`search` and `rangeSearch`) guarantee the most recent information, but they may take some time to discover these contents. The cache-based methods, in contrast, can provide results right away if they hit a suitable job in the cache. However, since the job has already been processed at a former point in time, its information might have become obsolete already. To adjust the caching behavior, `getMaxCacheTime` and `setMaxCacheTime` can be used. Old and orphaned jobs in the cache can be deleted with `cleanCache`, and `getJobCount` returns the number of jobs which reside in the cache.

The method `getAverageSearchTime` can be called to inquire the time recent search jobs needed in average to gather information in the peer-to-peer layer. This value is updated by `updateSearchTimeHistory`. How many jobs are taken into account to calculate this mean value can be adjusted with `getSearchTimeHistoryLength` and `setSearchTimeHistoryLength`.

Finally, `getJobStatus` allows an initiator to monitor the progress of a search. With `getSearchJob` the results of a `SearchJob` object can be fetched and returned to the caller. Besides all these methods which are mainly triggered by their counterparts in the `Core` class, `StorageService` implements the `SearchListener` interface of the P-Grid package. Similar how `P2PListener` in `MessageService` represents a call-back mechanism, `SearchListener` invokes the methods `searchStarted` and `searchFinished`, once a new search request enters and leaves the peer-to-peer layer. Between this start and end point, `newSearchResult`, `noSearchResult`, or `searchFailed` are invoked to indicate whether results haven been found, have not been found, or if something went wrong.

SearchJob. As already touched on explaining `StorageService`, instances of the `SearchJob` class represent the containers to store every piece of information that is related to a particular search request. Upon creating a new object, several parameters are captured. Among others, these are: the start time of the job, the unique identifier of the associated query, the key words, and the type of information which is requested. The values of these fields can be obtained calling get-ter methods like `getQueryGUID`, `getQueryString`, and `getQueryStringType`. Exactly these methods are invoked when `cachedSearch` or `cachedRangeSearch` seek for suitable jobs in the cache. The method `setResults` can be called to pass a search result to the

SearchJob object. These results can then be fetched with `getSearchResults`. The method `closeJob` is used when the search in the peer-to-peer layer has been finished. It stores the end time and changes the job's status the cache. With `getEndTime` and `getSearchTime` the important time-related values can be obtained. Finally, `getJobStatus` and `setJobStatus` are employed to determine a job's treatment in the cache. More on this is presented in 4.3.4.

Statistics. The last of APLICOOVER's major classes is `Statistics`. It implements the `Runnable` interface and hence provides a `run` method. This methods basically incorporates `Statistics`' processing logic. Once launched, it continuously captures and logs important performance parameters of APLICOOVER and P-Grid. Among others, it measures bandwidth consumption, message volume, and search times. For a complete listing of all parameters, please refer to the API documentation and the source code. The Statistics object is instantiated in class `Core`, handed over to a separate `Thread` object, and started. To stop monitoring, `shutdown` must be invoked which closes all logs files properly and frees the object.

Utils. Finally, the `Utils` class should be mentioned. It provides three methods which can't be attributed to any of the other classes. Their names are almost self-explaining: `extractGUID` and `extractName` return the GUID and name of a passed service description. A GUID, for instance, can be taken to refer to the particular description. The method `printResults` is mainly for testing purposes and prints a bundle of search results to text console.

Noticeably and worth mentioning is that APLICOOVER's central classes have been modeled as *Singletons* [Gam95]. This particularity mainly results from P-Grid's architecture which, for example, does not allow using more than one storage facility with the peer-to-peer communication channel. All data in P-Grid is routed through a single instance in its storage layer. There are several other single access points which quasi induce the design APLICOOVER possesses.

4.3 Integration

Drawing on selected examples, this section demonstrates how the presented classes interact and how APLICOOVER collaborates with the other modules of the middleware stack.

4.3.1 Bootstrap

APLICOOVER's initialization can be divided into two major steps. During the first one, the object is instantiated and registered in the RMI registry. This is triggered by the underlying agency and shown as the `new` call in Fig. 4.5. As always, the instantiation of the object causes the static `main` method to be invoked which handles exactly the described tasks. However, by just instantiating the object and registering it, none of its relevant services are ready to be used, since the module has not yet established a connection with the global overlay. This is done during the second step of the bootstrap procedure which is initiated by invoking `init`. This methods in turn calls `setProperties` to properly set all mandatory boot parameters. For example, the port is determined on which APLICOOVER listens for incoming messages from the peer-to-peer network. Having set these values, `sharedInstance` returns the handle `theP2PFactory` on the unique instance of the `PGridP2PFactory` class. Calling `createP2P` with the `properties` parameter, the factory objects builds the actual peer-to-peer instance `theP2PFacility`. Exactly this object represents the fundamental connection between the APLICOOVER layer and the peer-to-peer network. The facility object is used in the next step to create a new instance of the `StorageService` class. The instantiation of `theStorageService` links this object with `theP2PFacility` and P-Grid's storage layer, and registers it as the storage module to be used. Finally, the invocation of `registerType("service_description")` sets-up the storage module in that way that it is able to handle information of the given type, which is just plain text here. After the completion of these tasks, APLICOOVER's basic features are ready to be used.

During the subsequent steps, its enhanced functionalities are build-up. On the one hand, this is the object `theStatisticsService`. The objects `theP2PFacility` and `theStorageService` are passed to its constructor to register these entities which are about to be monitored. The reference to `theP2PFactory` object is needed for basic initialization issues. The created `Statistics` object is then handed over to a `Thread` object. The invocation of `start` triggers the `Statistics`' object `run` method, which henceforth captures various performance parameters of APLICOOVER and P-Grid and logs them to an XML file which can be used for various performance analysis. The second additional service represents `theMessageService` object of the `MessageService` class. Through a call of `addP2PListener`, this object becomes tied to `theP2PFacility` instance. From this time on, it serves as the access point for any kind of message traffic that is not related to `theStorageService`'s tasks.

Once all these preparations have been done, the actual bootstrap procedure in the sense of P-Grid can be initiated by invoking `bootstrapping`. That will cause the method `getOldNeighbors` to be called which returns the list of all proximate remote peers the local APLICOOVER instance has ever had a connection to. Each entry of this list is then taken and passed to the `join` methods which tries to re-establish that particular connection and registers the local instance with the peer-to-peer overlay network again. Since that might take some time, the handshake between peers is executed asynchronously, and `join` returns immediately. If it is the first time APLICOOVER is launched, the list of former neighbors does not exist yet. Then, it is attempted to contact a default bootstrap peer whose address has been determined in `setProperties`. If that fails too, the local APLICOOVER instance builds-up its own peer-to-peer system with just one member and starts listening for connection attempts of remote sites. This marks the end point of APLICOOVER's initialization, and a status flag, indicating whether everything went okay, is returned to the caller.

The shutdown process is not shown in the sequence diagram, but it is sketched quickly: Shutting APLICOOVER down is straightforward and basically happens in the opposite sequence as the bootstrap procedure. The invocation of `shutdown` in the `Core` class will trigger the corresponding methods to save the list of neighbor peers, and call each module's `shutdown` method, respectively. Thereby, APLICOOVER cuts all its links to the P-Grid tier, unregisters the references in the RMI Registry, and frees all associated objects.

4.3.2 Publish region profile

Once the initialization phase has been finished successfully, all of APLICOOVER's services can be employed without restriction. The next step should be to publish the region profile of the QuickLink group the APLICOOVER instance is running in. This profile is prepared and maintained by ServiceJuggler. As it in shown in Fig. 4.6, the latter invokes `publishRegionProfile` of the `Core` object as soon as the profile becomes available. However, since ServiceJuggler in its current implementation stage does not pass any parameters directly, the `Core` instance calls `RMIgetDominfo` to retrieve the profile. Basically, a profile is nothing more than a list where each entry represents a reduced description of an application service hosted by at least one agency of the local group. Reduced means that it only states the name of the service and how often it is available in the group, but not which particular agency / agencies host it. It also makes no sense at this point to publish port numbers, because each agency may run the service on a different port. The formal syntax of such a list entry looks like this:

```
service_description := service_name + '#' + service_cardinality
```

and is stored as a `String` object. To actually publish the profile, its records are passed to `publishService` one by one. The latter then calls `publishData` of the `StorageService` instance. This method wraps the description to a form that P-Grid understands and finally invokes `add(item)` which represents the point where handling of the item is directed to the P-Grid layer. The method returns immediately, since P-Grid processes this request asynchronously. It returns the global unique identifier (GUID) which identifies the item in the entire agent system. The GUIDs of all published items are stored in `publishService` until the entire profile has been published. Once this has been done, the bundle of GUIDs is returned to ServiceJuggler. In particular, the returned list comprises `String` objects of the form:

```
published_item := service_name + '_~_' + item_GUID.
```

Although ServiceJuggler does not draw on this information so far, it might be of great benefit for future versions, since GUIDs allow the exact identification of elements which then can be modified or deleted easily. GUID-based interaction strives

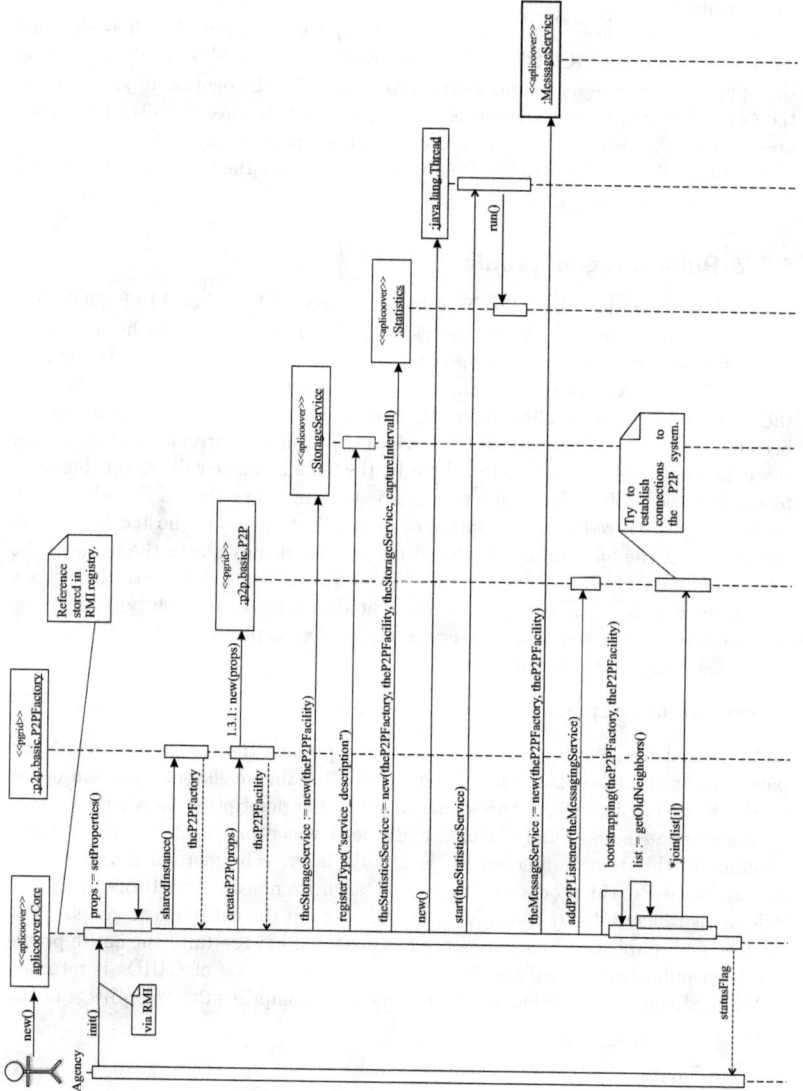

Figure 4.5: APLICOOVER's bootstrapping process.

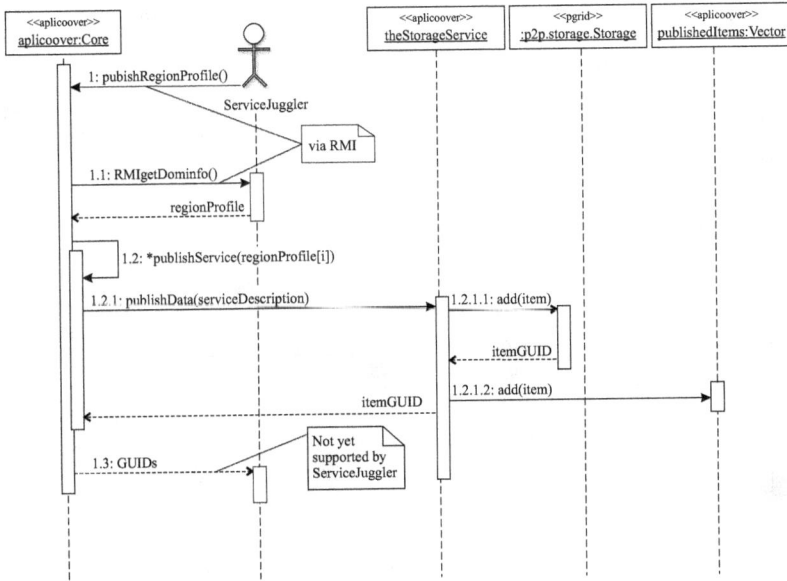

Figure 4.6: Publishing the region profile with APLICOOVER.

towards a fine-grained, accurate handling of published profiles. APLICOOVER already supports such interaction and its methods have GUID-ready signatures. Deleting of the entire profile can be realized with a call of `deleteRegionProfile`. Thereby, the vector `publishedItems` is run through and each entry is deleted in the global overlay. After the method returns, none of the published services can be discovered any more.

4.3.3 Search for service descriptions

Searching with APLICOOVER is executed in two steps. In the first one, a new search job is launched and seeks to discover the requested information in the peer-to-peer layer. This is shown in Fig. 4.7. The second step represents the information retrieval, as it is shown in Fig. 4.8. The search starts with an invocation of the

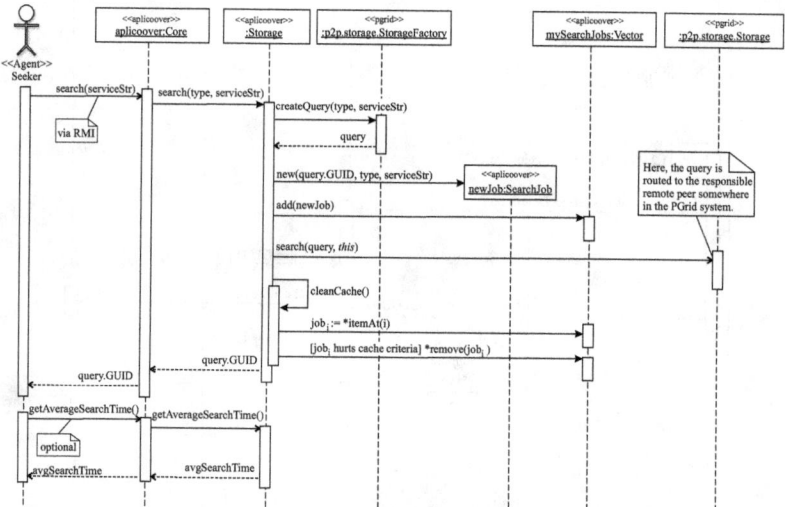

Figure 4.7: Searching for application services using APLICOOVER.

`search` method of the `Core` instance. Its single parameter is the keyword which describes the information the caller is looking for. Generally, this would be the name for a service given as a `String` object. `Core`'s `search` method formats the keyword and passes it to the `search` method the `StorageService` instance. There, the request is wrapped with a type object (see section 4.2) to a query object by engaging `createQuery` of the `PGridStorageFactory` instance. The characterizing parameters of the query object are then passed to the constructor of the `SearchJob` class. Henceforth, this object is used to monitor the course of the search and accept any results received from the peer-to-peer overlay. As every new `SearchJob` object, it is put in the data structure which is called cache. Once this has been done, the `search` method of the P-Grid storage layer is invoked and handed over the query object. With this invocation, the request is routed to the responsible peer somewhere in the global agent system and the parameter `this` registers the `StorageService` instance of the local APLICOOVER instance as the receiver for any results. Therefor it implements the `SearchListener` interface which contains the necessary call-back methods.

Having finished the handshake with the P-Grid tier, `StorageService`'s `search` method calls `cleanCache` which in turn scans `mySearchJobs` for obsolete and orphaned `SearchJob` objects. See section 4.3.4 for further details. Finally, the GUID of the query object is returned to the caller so that it can be used to inquire the status of the search and retrieve the results. It is optional to invoke `getAverageSearchTime`. Doing so will return the average search time of the recent queries. This gives the caller a rough idea when to expect results.

The second step of a search request with APLICOOVER starts basically as soon as a query result is returned from the peer-to-peer overlay. In particular, the unique `p2p.basic.P2P` object `theP2PFacility`, which had been instantiated when APLICOOVER was launched, receives such messages. Whether it was a successful search or not, the corresponding call-back method of the `StorageService` instance is invoked. Assuming it was a successful search, `newSearchResult` is triggered. Besides the result set, the GUID of the initial query is passed to identify the corresponding `SearchJob` object in cache. With this GUID `mySearchJobs` is scanned for the associated job the received result belongs to. Let this job be `aPendingJob`. Once this job has been found, `newSearchResult` calls `setResult` to store the result set with this `SearchJob` object. Then, the job's status is adjusted by calling `setStatus(SUCC_FINISHED)`. Which consequences arise with this change regarding the object's treatment in the cache is explained in section 4.3.4. Finally, the job is closed and `updateSearchTimeHistory` is called. The latter inquires the job's total search time and incorporates it in the value which can be obtained using `getAverageSearchTime`. Once the `newSearchResult` method returns, the results are stored within the associated `SearchJob` object and its status is adjusted properly.

The agent which initiated the search is able to monitor the course of the search at any time employing `getQueryStatus`. Thereby, the query's GUID serves once more as a signature to identify the job in cache. Assuming that this job is exactly `aPendingSearchJob`, the status `SUCC_FINISHED` is returned to the agent what signalizes that pending search results are ready to be retrieved. There are several other values that describe the course of a search. For a complete list, please consult section 4.3.4 and APLICOOVER's API documentation.

The result set can be retrieved invoking `getSearchResults` of the `Core` instance. Drawing on the GUID, the job is identified and its results are returned. They are converted to a flexible format which allows better processing for the agent. Each entry of the `formatedResults` object has the structure:

```
found_service := service_name + '#' + cardinality + '@' + ip
```

In other words, the name and the cardinality of the service is returned, as well as the IP address of the remote APLICOOVER instance which has published this item as part of the remote region profile. That means the remote agency, hosting the APLICOOVER instance, serves as the gateway for any mobile agent that wishes to use an identified service in that region.

An invocation of the rangeSearch method of the Core instance basically results in

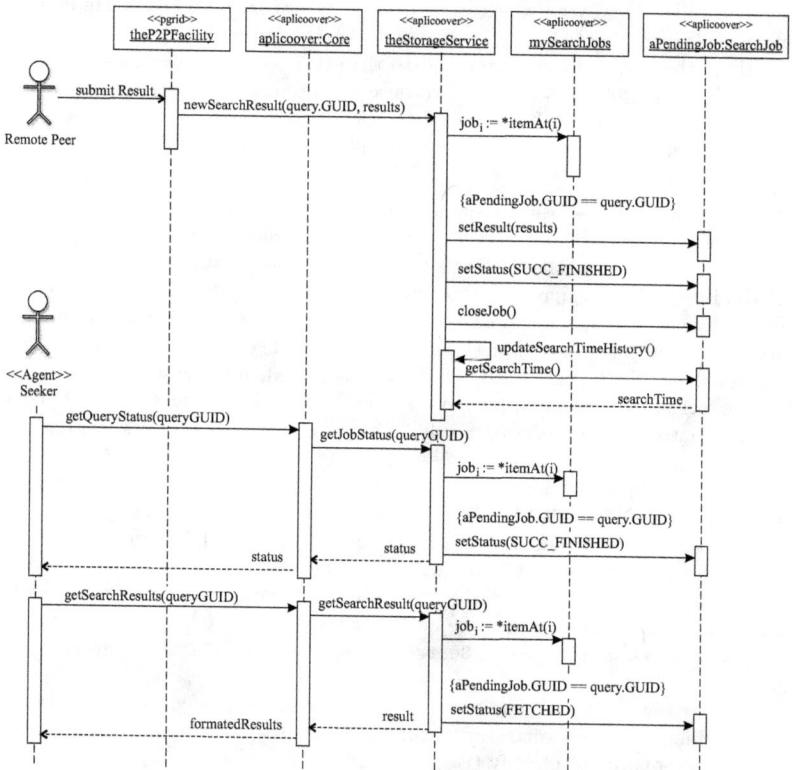

Figure 4.8: Retrieving the search results gathered by APLICOOVER.

a same interactions between the classes. The only difference is that two keywords are passed as parameters which determine the lower and the upper bound of the search space which is going to be scanned for matching items.

Even the cached methods run similar. They only differ from the standard methods by including the cache to seek for suitable results. An invocation of cachedRangeSearch will first scan `mySearchJobs` for jobs which have the same keywords, the same type, and are already finished successfully. If a suitable job is found, the job's GUID is returned to the caller and it can fetch the results. However, if the cache contains no similar job, a flag determines what happens further: If it is not set, search ends here. In the other case, a new `SearchJob` object is created and a regular search is initiated as it is shown above.

4.3.4 The cache protocol

The technical detail that should be addressed here last is APLICOOVER's cache protocol. The cache is a thread-safe vector which is used to store `SearchJob` objects. The cache-based methods of the `Core` object scan this cache for suitable information before they escalate to the global level to continue searching. The cache is maintained, using a cache protocol which assigns different states to the search jobs. Fig. 4.9 shows which states are possible and how they are related. When a new `SearchJob` object is created, its status is set to *unknown*. As soon as the associated query is passed to the P-Grid layer, the actual search starts and the call-back method `searchStarted` of APLICOOVER's `StorageService` instance is invoked. Thereby, the status of the job is changed to *started* and the start time is stored with the `SearchJob` object. Entering this state represents the hand-over of control to the P-Grid layer. The job remains in this state until a reply from the peer-to-peer layer has been received and control is granted back to APLICOOVER. If the search succeeded, `newSearchResult` is triggered, the result set is stored with the `SearchJob` object, and its status is changed to *successful*. If no results could be found in the peer-to-peer layer or if technical problems caused any failure while routing the request, `noSearchResults` and `searchFailed` are called, respectively. They switch the status to `unsuccessful` or `failed`. If a job has been assigned either one of the latter states and if the initiator of the request has already inquired the status using the `getStatus` method, it is assumed that the job is of no interest anymore. Hence, its status is switched to *no result & status inquired*, and the job is going to be deleted immediately with the next invocation of `cleanCache`. The same happens to an unsuccessful or canceled job whose status has not been inquired yet, but which has reached the maximum of time a job is allowed to reside in cache.

In case of a successful search, the job's status is changed to *result & status inquired*. From this status or the previous one already, a cached-based method adjusts this value to *in cache and of interest* if the job represents a cache-hit, i.e. contains exactly the information the new request is seeking for. Alternatively, its results can be retrieved and its state is switched to `fetched`. From either one of theses states, the job is deleted when it resides too long in the cache. Consequently, the cache protocol tries to keep as much jobs in the cache to higher the probability of cache-hits. At the same time, old and orphaned jobs are deleted to keep the data volume as low as possible and expedite scan processes.

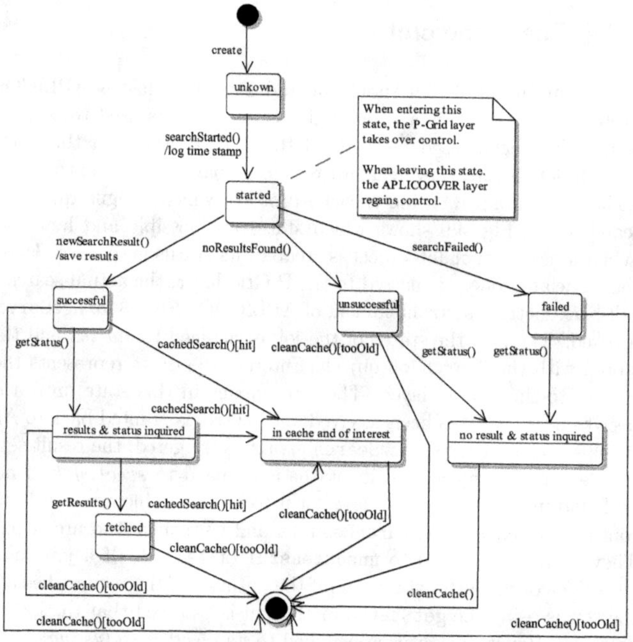

Figure 4.9: APLICOOVER's cache protocol.

4.4 Tests

To test APLICOOVER's features, it underwent a comprehensive test series that verified that all its services work as expected. The test environment was made-up by four PCs (all above 512 MB RAM and 2GHz) which were connected through a 100 Mbit full-duplex Ethernet network. As operating systems, openSuSE Linux 10.1 and Microsoft Windows XP Professional have been used. On all devices, APLCIOOVER was compiled with SUN JDK1.5.0_06 and launched from Borland JBuilder Enterprise 2005.

For the functional test series, a dummy instances of ServiceJuggler has been used. It provides the same functionality to properly interact with APLICOOVER, but does not rely on actual service data of a QuickLink group. The utilized service descriptions were simply generated. The underlying agency as well as the agents were also emulated by a test protocol. It was used to start / stop APLICOOVER and trigger the methods agents would normally use. It particular, it had the following structure:

1. Create an APLICOOVER instance and register it in the RMI registry

2. Initialize APLICOOVER

3. Publish the region profile

4. Start a query that should return all published descriptions

5. Start a query that should return exactly one published description

6. Modify one description

7. Delete one description

8. Start a query to search for the updated item

9. Start a query to search for the deleted item

10. Start a range search

11. Start a cached search (cache-hit)

12. Start a cached range search (cache-miss, globally continued)

13. Shutdown the APLICOOVER component

After the first two step of this sequence, the protocol pauses activity for three minutes to allow P-Grid entering the replication phase and to make sure that every peer elaborates its path and builds-up its routing table. During the the third step finally, APLICOOVER fetches the region profile from the dummy ServiceJuggler via a RMI call and publishes the information. After that, the test protocol pauses again for three minutes to assure that P-Grid has time to initiate the bootstrapping phase. Then the remaining part of the protocol was processed sequentially.

Each one of the invocations resulted in the expected action. Especially all search methods discovered exactly the data they should find. For the complete results, please refer to A. With these results, APLICOOVER has passed the functional test series without any problems and hence provides the required functionality to the full extend.

After these first tests have been successfully accomplished, attention was directed to the waiting states after the first and third step of the protocol. For the set of all four computers, the periods could be reduced to approximately 2 minutes, for two devices even down to 100 seconds. Waiting states below these thresholds caused in odd behaviors. For example, during step 4 of the test sequence not all service descriptions were found. This was most likely caused by lacunar routing tables which could not be fully elaborated during the short amount of time between the initialization and publication. If an APLICOOVER instance was paused immediately after step 3, its routing table sometimes did not contain all peers and/or a peer's path was not set yet. It is obvious that such issues might cause routing problems which explain the missing results.

Sometimes however, one could also observe that items were listed twice during steps 4 and 5 if the waiting states were below the threshold values. This phenomenon could be caused by issues that arose out of an interrupted replication phase in the P-Grid layer. Most likely, peers were just about to replicate their data when searching already started, but did not have the time to adjust their routing tables properly. For example, it might have been that a peer was still listing its (just recently established) replica in the reference part of the routing table, but did not have the time to add the entry in the replicas section. That might result in odd routing behaviors for queries. Since the replica and the original peer possess the same path at that moment, but the replica is not actually counted as replica yet, a query might be routed to two peers with exactly the same data which in turn would explain double results for the search attempts. Unfortunately, these kinds of side-effects were hardly to reproduce and every time the routing tables of the peers were examined no irregularities could be found which would back-up this hypothesis.

Driven to the extreme, i.e. no waiting states were used after the initial step and the publication, P-Grid even crashed. A trace of the exception showed that the local P-Grid database needs some time for initialization issues, too. If one grants the threshold values, none of these odd behaviors have ever been observed. To which extend these waiting states must be adjusted as soon as bigger peer populations are interconnected could not be determined by this work, since the limited number of devices the tests were executed on did not allow any meaningful analysis in this direction. Hence, such tests would be more than welcome to find the delimiting properties and identify possible bottlenecks.

4.5 Evaluation and outlook

The functional tests with APLICOOVER demonstrated that the module works properly and is ready to be integrated into the QuickLinkNet environment. The next step should be to impel a comprehensive series of tests which assess APLI-COOVER's performance for realistic system sizes. Only such tests then allow a final evaluation if the module can stand the expectations which derive from the given constraints. These tests should try to enlighten the following characteristics:

- How long does it take to discover a newly published item for the first time?

- How long does it take until a newly published item is discovered by all peers?

- Which influence does the system size have on the waiting states, as they are mentioned in section 4.5.1?

- Does the size alone determine the lengths of the pauses, or does one also have to consider the system's dynamics?

- How long does it take until increasingly large systems cease oscillating, i.e. when does every peer have been assigned a path after system boot-up?

- To which extend does the communication overhead increase to establish and maintain the overlay structure depending on the system's size / dynamics?

- How much does the communication volume depend on the number of published items / the key distribution?

- Is the overlay structure, especially the depth of the tree, driven by the system's size, its dynamics, or the key distribution?

- How do search times vary with respect to these parameters?

- Does the volume of routing traffic for search requests depend on these characteristics?

The test series also revealed some points of improvement which might be considered for the next implementation stage of APLICOOVER. For example:

- Although APLICOOVER's search methods work fine and efficiently, the way how searching is done, has still room for improvement. Originally, the separation between initiating a search request and retrieving its results was done to correspond to an agent's autonomy. Thus, the current design allows agents to start a search request and to freely decide when to inquire the status or retrieve the results. All action is triggered by the agent, whereas APLICOOVER takes a rather passive role. However, a call-back mechanism that notifies the initiator of a request as soon as a search result has been received could be very helpful. If this method should be considered, an asynchronous communication channel would be desirable. It could be used to simply ping the agent, but not to force it to take over a result set immediately. Instead, it should be still up to the agent when to fetch a result. The call-back mechanism would basically allow retrieving results as soon as they become available. Currently, the utilization of `getStatus` method resembles rather a polling mechanism.

- Similar how ServiceJuggler allows the migration of its directory service, APLI-COOVER should implement some mechanism that would allow to switch its site of execution without the whole initialization procedure. Currently, shutting APLICOOVER down on one agency and relaunching it on another, requires the whole region profile to be deleted and republished again. With regard to the dynamics of a region, this results in additional administration overhead and network traffic that could be avoided to a certain degree. It seems easily possible to shutdown APLICOOVER, without deleting the region profile, but instead updating it as soon as the module has been relaunched on a new agency. Basically, only the IP addresses of the published items must be adjusted, since the new agency then serves as the new gateway to the region. This strategy seems to require less effort and results in fewer messages that a regular startup.

- Some of APLICOOVER's methods could work more efficiently if subtasks are handed over to separate threads. For example, the bootstrapping procedure in `init` of the `Core` class could be done in background. The extra thread could read the list of former neighbor peers and try to contact them sequentially. In much the same way, the `cleanCache` method of the `StorageService` class could run continuously in background. That would speed-up the search methods which currently scans the cache each time a new search request is initiated.

- On an even lower level, APLICOOVER's design must probably be questioned again. Although P-Grid's single-point-of-access architecture virtually forces any client module to follow this design, several bottlenecks could probably be circumvented. Using dedicated queues for particular tasks could be a worthwhile possibility. Such queues could be processed by separate threads which pass the tasks the corresponding the P-Grid interfaces. Although this would have no direct influence on P-Grid's processing speed, it would allow APLI-COOVER to buffer requests as serve more clients, i.e. agents. However, even such techniques cannot eliminate all of P-Grid's weak points. For example, the instantiation of a new query object happens via a static method of the `PGridStorageFactory` class. Exactly these points, where APLICOOVER touches the P-Gird layer, show that the employed systems are motivated from different directions: Whereas APLICOOVER demands for an agent-friendly design, P-Grid at some points seems still heavily influenced by the file-sharing sector. There, one peer represents exactly one (human) actor. Hence, static methods and single points of access are sufficient, since only one actor interacts with the system. With P-Grid's utilization in APLICOOVER, multiple actors—like ServiceJuggler, agents, etc.—must be served. This results in the emergence of bottlenecks and hot-spots which, in extreme situations, might bring the system to a point where it cannot be employed as required.

- On a very abstract level, which does not relate to APLICOOVER alone, the problem of keyword-based searching must be questioned. In particular this means application services within an agent system can only be discovered if the seeker has at least a rough idea of the service's name it is looking for. Neither APLICOOVER nor P-Grid currently offer advanced wild-card mechanisms or even semantic techniques to search for descriptions. For example a request like search('ServiceXYZ') would not find a service which has been published as 'myServiceXYZ'. Here it becomes obvious that string-based search techniques become challenged quickly and semantic solutions, like the utilization of ontologies, should generally be favored.

5 Conclusion

This book addressed mobile agent systems and the fundamental challenges that arise with proactive, autonomous, service-oriented migration strategies. Chapter 1 showed that there are already well-elaborated ideas, like virtual maps, route planners, and migration optimizers which strive for the realization of agent systems of the second maturity level. It has further been shown that the underlaying infrastructure of current available agent toolkits usually possesses central elements. These central entities do not only represent bottlenecks, single-points-of-failure, and hot-spots for any kind of activity, but fundamentally contradict the peer character of agencies. Consequently, this book tried to sketch how agent systems can be understand as peer-to-peer systems where, beyond the agents themselves, the underlaying agencies too are treated as peers in the overall system.

It was finally postulated that peer-to-peer technology can serve as a suitable means to interconnect agencies to a coherent agent system. This liaison seemed very promising since both technologies have similar traits of character and are able to complement each other in a meaningful way. However, the hard constraints of mobile agent technology itself and its targeted application domain leave only small corridors and disqualify many peer-to-peer implementations right away. After a general introduction in chapter 2, the next chapter presented three selected peer-to-peer implementations which have been assessed according to the required functionality and performance. Thereby, P-Grid has been identified as a promising candidate to prove the feasibility of the envisioned concept. Its characteristics seemed to match the constraints of an agent system perfectly, it is well documented, and it is freely available.

In chapter 4, the prototype APLICOOVER was introduced which has been developed as part of this book. On basis of P-Grid, it represents a middleware component which interconnects agencies / groups of agencies to a large-scale agent system. It also facilitates the information management on this global level. In particular it can be used to publish, update, delete, and search for service descriptions, as they are required by mobile agents to allow service-oriented navigation in the agent system. Drawn on Erfurth's idea [Erf04] it could be shown that APLICOOVER is perfectly suited to complement already existing modules of the software stack to facilitate the creation of virtual maps of the agent system. These maps in turn can be used to serve as means of orientation for mobile agents and thus allow to identify the services they require to fulfill their tasks. Additionally, with APLICOOVER it becomes possible to discover information that has been published anywhere in the agent system. Current implementations usually limit such search requests to the proximate neighborhood of an agency.

The test series, assessing APLICOOVER's functionality, showed that the component works as expected and all features are ready to be used. This proves that mobile agent technology can highly benefit of peer-to-peer technology to build-up a global infrastructure of an agent system. Even more, peer-to-peer technology seems to perfectly correspond to agent technology's core. Bottom-up networking of distributed entities is exactly what agent technology propagates and demands for. Unfortunately, due to the limited testing environment, the performance tests could not fully answer if APLICOOVER can actually handle the challenges that arise with highly-dynamic large-scale agent systems. Consequently, the next step should be to assess its performance with a comprehensive test series and realistic system sizes. A promising platform to run such tests on could be PlanetLab [Pet06].

A Protocol of the functional tests

This listing shows the sequence of invocations that were used to test APLICOOVER's
functionality. Please refer to the source code for the complete version.

```
...
adminStub = (remote.RemoteAdministration) registry.lookup("APLICOOVER");
standardStub = (remote.RemoteServices) registry.lookup("APLICOOVER");
...

//Bootstrap
adminStub.init();
...

//Publish region profile
String[] bundleDescription = adminStub.publishRegionProfile();
...

//Searching
String queryGUID = standardStub.query("ApplicationService");
//Searching
queryGUID = standardStub.query("ApplicationService1_192.168.10.1");
...

//Modify first service
String guidOfModifiedService = utils.Utility.extractGUID(bundleDescription[0]);
standardStub.updateService(guidOfModifiedService, "ModifiedApplicationService");
...

//Delete second service
String guidOfDeletedService = utils.Utility.extractGUID(bundleDescription[1]);
standardStub.deleteService(guidOfDeletedService);
...

//Searching for modified service
queryGUID = standardStub.query("Modified");
...

//Searching for deleted service
queryGUID = standardStub.query("ApplicationService2_192.168.10.1");
...

//Range search
queryGUID = standardStub.rangeQuery("A","M");
...

//Cached search, non globally, cache-hit
//to-be-searched information is deleted in the p2p layer first
//to make sure the results come from cache
standardStub.deleteService(guidOfModifiedService);
...
queryGUID = standardStub.cachedQuery("Modified", false);
...

//Cached range search - globally
queryGUID = standardStub.cachedRangeQuery("Ap","Mo",true);
...

//shutdown
adminStub.shutdown();
```

This is the condensed output of the testbed:

```
Launching bootstrap...done.
Publishing service descriptions......done.
...

Searching for 'ApplicationService'...
ApplicationService1_192.168.10.1_#1@192.168.10.1
ApplicationService2_192.168.10.1_#2@192.168.10.1
ApplicationService3_192.168.10.1_#3@192.168.10.1
ApplicationService4_192.168.10.1_#4@192.168.10.1
ApplicationService2_192.168.10.15_#2@192.168.10.15
ApplicationService1_192.168.10.15_#1@192.168.10.15
ApplicationService3_192.168.10.15_#3@192.168.10.15
ApplicationService4_192.168.10.15_#4@192.168.10.15
ApplicationService3_192.168.10.20_#3@192.168.10.20
ApplicationService4_192.168.10.20_#4@192.168.10.20
ApplicationService2_192.168.10.20_#2@192.168.10.20
ApplicationService1_192.168.10.20_#1@192.168.10.20
ApplicationService1_192.168.10.16_#1@192.168.10.16
ApplicationService3_192.168.10.16_#3@192.168.10.16
ApplicationService4_192.168.10.16_#4@192.168.10.16
ApplicationService2_192.168.10.16_#2@192.168.10.16

Searching for 'ApplicationService1_192.168.10.1'...
ApplicationService1_192.168.10.1_#1@192.168.10.1
ApplicationService1_192.168.10.16_#1@192.168.10.16
ApplicationService1_192.168.10.15_#1@192.168.10.15

Updating first published service...
Deleting second published service...
Searching for 'Modified'...
ModifiedApplicationService@192.168.10.1
ModifiedApplicationService@192.168.10.16
ModifiedApplicationService@192.168.10.20
ModifiedApplicationService@192.168.10.15

Searching for deleted service...

Range search for 'A'-'M'...
ApplicationService3_192.168.10.1_#3@192.168.10.1
ApplicationService4_192.168.10.1_#4@192.168.10.1
ModifiedApplicationService@192.168.10.1
ModifiedApplicationService@192.168.10.16
ModifiedApplicationService@192.168.10.20
ModifiedApplicationService@192.168.10.15
ApplicationService3_192.168.10.15_#3@192.168.10.15
ApplicationService4_192.168.10.15_#4@192.168.10.15
ApplicationService4_192.168.10.16_#4@192.168.10.16
ApplicationService4_192.168.10.20_#4@192.168.10.20
ApplicationService3_192.168.10.20_#3@192.168.10.20
ApplicationService3_192.168.10.16_#3@192.168.10.16

Deleting modifed service...
Cached search (not globally cont'd) for 'Modified'...
Found in cache:
ModifiedApplicationService@192.168.10.1
ModifiedApplicationService@192.168.10.16
ModifiedApplicationService@192.168.10.20
ModifiedApplicationService@192.168.10.15

Cached range search (globally cont'd) for 'Ap'-'Mo'...
Not found in cache but in the p2p layer:
ApplicationService3_192.168.10.1_#3@192.168.10.1
ApplicationService4_192.168.10.1_#4@192.168.10.1
ModifiedApplicationService@192.168.10.1
ModifiedApplicationService@192.168.10.16
ModifiedApplicationService@192.168.10.20
ModifiedApplicationService@192.168.10.15
ApplicationService4_192.168.10.16_#4@192.168.10.16
```

97

```
ApplicationService4_192.168.10.20_#4@192.168.10.20
ApplicationService3_192.168.10.16_#3@192.168.10.16
ApplicationService3_192.168.10.20_#3@192.168.10.20
ApplicationService3_192.168.10.15_#3@192.168.10.15
ApplicationService4_192.168.10.15_#4@192.168.10.15

Shutting down APLICOOVER instance...
```

Bibliography

[Abe01] Aberer, K.: *P-Grid: A Self-Organizing Access Structure for P2P Information Systems*, in *CoopIS '01: Proceedings of the 9th International Conference on Cooperative Information Systems*, Springer-Verlag, London, UK, 2001, S. 179–194.

[Abe02] Aberer, K.; Punceva, M.; Hauswirth, M.; Schmidt, R.: *Improving Data Access in P2P Systems*, IEEE Internet Computing, Bd. 6, Nr. 1, 2002, S. 58–67.

[Abe03a] Aberer, K.; Cudré-Mauroux, P.; Datta, A.; Despotovic, Z.; Hauswirth, M.; Punceva, M.; Schmidt, R.; Wu, J.: *Advanced Peer-to-Peer Networking: The P-Grid System and its Applications*, PIK Journal - Praxis der Informationsverarbeitung und Kommunikation, Special Issue on P2P Systems, 2003.

[Abe03b] Aberer, K.; Datta, A.; Hauswirth, M.: *The Quest for Balancing Peer Load in Structured Peer-to-Peer Systems*, 2003.

[Abe03c] Aberer, K.; Punceva, M.: *Efficient Search in Structured Peer-to-Peer Systems: Binary vs. k-ary Unbalanced Tree Structures*, 2003.

[Abe04] Aberer, K.; Datta, A.; Hauswirth, M.: *Multifaceted Simultaneous Load Balancing in DHT-based P2P systems: A new game with old balls and bins*, 2004.

[Abe05] Aberer, K.; Datta, A.; Hauswirth, M.; Schmidt, R.: *Indexing data-oriented overlay networks*, in *VLDB '05: Proceedings of the 31st international conference on Very large data bases*, VLDB Endowment, 2005, S. 685–696.

[Acq02] Acquisti, A.; Sierhuis, M. et al.: *Agent Based Modeling of Collaboration and Work Practices Onboard the International Space Station*, in *11th Computer-Generated Forces and Behavior Representation Conference, Orlando, Fl*, 2002.

[AT04] Androutsellis-Theotokis, S.; Spinellis, D.: *A survey of peer-to-peer content distribution technologies*, ACM Comput. Surv., Bd. 36, Nr. 4, 2004, S. 335–371.

[Bar99] Barabasi, A.-L.; Albert, R.: *Emergence of scaling in random networks*, Science, Bd. 286, 1999, S. 509.

[Bau00] Baumer, C.; Breugst, M.; Choy, S.; Magedanz, T.: *Grasshopper: a universal agent platform based on OMG MASIF and FIPA standards*, 2000.

[Bel00] Bellavista, P.; Corradi, A.; Stefanelli, C.: *A mobile agent infrastructure for the mobility support*, in *SAC '00: Proceedings of the 2000 ACM symposium on Applied computing*, ACM Press, New York, NY, USA, 2000, S. 539–545.

[Bel01a] Bellifemine, F.; Poggi, A.; Rimassa, G.: *Developing Multi-agent Systems with JADE*, in *ATAL '00: Proceedings of the 7th International Workshop on Intelligent Agents VII. Agent Theories Architectures and Languages*, Springer-Verlag, London, UK, 2001, S. 89–103.

[Bel01b] Bellifemine, F.; Poggi, A.; Rimassa, G.: *JADE: a FIPA2000 compliant agent development environment*, in *AGENTS '01: Proceedings of the fifth international conference on Autonomous agents*, ACM Press, New York, NY, USA, 2001, S. 216–217.

[Ben96] Benerecetti, M.; Cimatti, A.; Giunchiglia, E.; Giunchiglia, F.; Serafini, L.: *Formal Specification of Beliefs in Multi-Agent Systems.*, in Müller et al. [Mül97], S. 117–130.

[Boo04] Booth, D.; Haas, H.; McCabe, F.; Newcomer, E.; Champion, M.; Ferris, C.; Orchard, D.: *Web Services Architecture*, W3C Working Group Note 11, 2004, http://www.w3.org/TR/ws-arch/.

[Bra97] Bradshaw, J. M.: *An Introduction to Software Agents*, in Bradshaw, J. M. (Hrsg.): *Software Agents*, AAAI Press / The MIT Press, 1997, S. 3–46.

[Bra04] Braun, P.; Muller, I.; Geisenhainer, S.; Schau, V.; Rossak, W.: *A Service-oriented Software Architecture for Mobile Agent Toolkits*, ecbs, Bd. 00, 2004, S. 550.

[Bra05] Braun, P.; R. Rossak, W.: *Mobile Agents-Basic Concept, Mobility Models, and the Tracy Toolkit*, Morgan Kaufmann Publishers, 2005.

[Bre01] Breidenbach, S.: *Peer-to-Peer Potential*, http://www.networkworld.com/research/2001/0730feat.html, 2001.

[Bro02] Broekstra, J.; Kampman, A.; van Harmelen, F.: *Sesame: A generic architecture for storing and querying rdf and rdf schema*, 2002.

[Buc04] Buccafurri, F.; Lax, G.: *TLS: A Tree-Based DHT Lookup Service for Highly Dynamic Networks.*, in *CoopIS/DOA/ODBASE (1)*, 2004, S. 563–580.

[Cla02] Clarke, I.; Hong, T. W.; Miller, S. G.; Sandberg, O.; Wiley, B.: *Protecting Free Expression Online with Freenet*, IEEE Internet Computing, Bd. 6, Nr. 1, 2002, S. 40–49.

[Con00] Content, F.; XC, L.; www, A.; f specs, ; f doc, : *FIPA SL Content Language Specification*, 2000.

[Dab01] Dabek, F.; Brunskill, E.; Kaashoek, M. F.; Karger, D.; Morris, R.; Stoica, I.; Balakrishnan, H.: *Building Peer-to-Peer Systems with Chord, a Distributed Lookup Service*, 2001, S. 81–86.

[Dat03] Datta, A.; Hauswirth, M.; Aberer, K.: *Updates in Highly Unreliable, Replicated Peer-to-Peer Systems*, in *In the proceedings of the 23rd International Conference on Distributed Computing Systems, ICDCS2003 (to appear)*, 2003.

Bibliography

[Dig03]   Dignum, F.: *Web Services and Software Agents.*, in *iiWAS*, 2003.

[Doe05]   Doehler, A.; Erfurth, C.; Rossak, W.:   *A Framework of Autonomous and Self-adaptable Middleware Services to Support Mobile Agents in Dynamic Networks*, GESTS International Transactions on Computer Science and Engineering, Bd. 19, Nr. 1, 2005, S. 109–122.

[dS04]   da Silva, V. T.; de Lucena, C. J. P.: *From a Conceptual Framework for Agents and Objects to a Multi-Agent System Modeling Language.*, Autonomous Agents and Multi-Agent Systems, Bd. 9, Nr. 1-2, 2004, S. 145–189.

[ea04]   et al., P. H.: *Bibster — a semantics-based bibliographic peer-to-peer system*, in *Proc. of the 3rd Int. Semantic Web Conference, Hiroshima, Japan*, 2004.

[ECM02]   ECMA, : *ECMA-335: Common Language Infrastructure (CLI)*, Second. Ausg., Dez. 2002.

[Erf04]   Erfurth, C.: *Proaktive autonome Navigation fr mobile Agenten.*, PhD thesis, Friedrich-Schiller Universität, Insititut für Informatik, 07743 Jena, Germany, Juli 2004.

[Far00]   Farjami, P.; Gerg, C.; Bell, F.: *Advanced service provisioning based on mobile agents*, Computer Communications, Special Issue: Mobile Software Agents for Telecommunication Applications, Bd. 23, Apr 2000, S. 754–760.

[Fei05]   Feier, C.; Roman, D.; Polleres, A.; Domingue, J.; Stollberg, M.; Fensel, D.: *Towards Intelligent web Services: Web Service Modeling Ontology (WSMO)*, in *In Proceedings of the International Conference on Intelligent Computing (ICIC)*, Hefei, China, August 2005.

[Fin94]   Finin, T.; Fritzson, R.; McKay, D.; McEntire, R.: *KQML as an Agent Communication Language*, in Adam, N.; Bhargava, B.; Yesha, Y. (Hrsg.): *Proceedings of the 3rd International Conference on Information and Knowledge Management (CIKM'94)*, ACM Press, Gaithersburg, MD, USA, 1994, S. 456–463.

[fip04]   *FIPA Agent Management Specification*,   `http://www.fipa.org/specs/fipa00023/`, 2004.

[Fra96]   Franklin, S.; Graesser, A. C.: *Is it an Agent, or Just a Program?: A Taxonomy for Autonomous Agents.*, in Müller et al. [Mül97], S. 21–35.

[Gam95]   Gamma, E.; Helm, R.; Johnson, R.: *Design Patterns. Elements of Reusable Object-Oriented Software*, Addison-Wesley Professional Computing Series, Addison-Wesley, 1995, GAM e 95:1 1.Ex.

[Gle04]   Gleichmann, N.: *Avatare (virtuelle Figuren) zur Interaktion mit Menschen*, Juli 2004.

[gnu04]   *The Gnutella Developer Forum*, `http://rfc-gnutella.sourceforge.net`, 2004.

[Gon01]   Gong, L.: *Project JXTA: A Technology Overview*, 2001.

[Hau05a] Hauswirth, M.; Datta, A.; Aberer, K.: *Handling Identity in Peer-to-Peer Systems*, 2005.

[Hau05b] Hauswirth, M.; Dustdar, S.: *Peer-to-Peer: Grundlagen und Architektur*, Datenbank-Spektrum, Bd. 5, Nr. 13, 2005, S. 5–13.

[HCB77] Henry C. Baker, J.; Hewitt, C.: *The incremental garbage collection of processes*, in *Proceedings of the 1977 symposium on Artificial intelligence and programming languages*, ACM Press, New York, NY, USA, 1977, S. 55–59.

[He03] He, M.; Jennings, N. R.; Leung, H.-F.: *On Agent-Mediated Electronic Commerce*, IEEE Transactions on Knowledge and Data Engineering, Bd. 15, Nr. 4, 2003, S. 985–1003.

[Her03] Hericko, M.; Juric, M. B.; Rozman, I.; Beloglavec, S.; Zivkovic, A.: *Object serialization analysis and comparison in Java and .NET*, SIGPLAN Not., Bd. 38, Nr. 8, 2003, S. 44–54.

[Hic98] Hicks, C.; Hines, S. A.; Harvey, D.; McLeay, F. J.; Christensen, K.: *An Agent Based Model of Supply Chains.*, in *ESM*, 1998, S. 609–613.

[Huh02] Huhns, M. N.; Stephens, L. M.; Ivezic, N.: *Automating supply-chain management*, in *AAMAS '02: Proceedings of the first international joint conference on Autonomous agents and multiagent systems*, ACM Press, New York, NY, USA, 2002, S. 1017–1024.

[IGD06] IGD, F.: *SeMoA - Secure Mobile Agents*, http://www.semoa.org, 2006.

[iiH05] ichiro itojun Hagino, J.: *IPv6 network programming*, 2005.

[Jab03] Jablonski, S.; Meiler, C.; Petrov, I.: *Web-Services und Semantic Web.*, HMD - Praxis Wirtschaftsinform., Bd. 234, 2003.

[Jon98] Jonker, C. M.; Treur, J.: *Agent-based simulation of animal behaviour*, in *248*, Centrum voor Wiskunde en Informatica (CWI), ISSN 1386-369X, 31 1998, S. 33.

[Kar00] Karp, R. M.; Schindelhauer, C.; Shenker, S.; Vocking, B.: *Randomized Rumor Spreading*, in *IEEE Symposium on Foundations of Computer Science*, 2000, S. 565–574.

[Kel02] Keleher, P.; Bhattacharjee, B.; Silaghi, B.: *Are virtualized overlay networks too much of a good thing*, 2002.

[Kra97] Kramer, J.: *Agent Based Personalized Information Retrieval*, 1997.

[Kum01] Kumar, K.: *Technology for supporting supply chain management: introduction*, Commun. ACM, Bd. 44, Nr. 6, 2001, S. 58–61.

[Küs03] Küster, M. W.: *Web-Services - Versprechen und Realität.*, HMD - Praxis Wirtschaftsinform., Bd. 234, 2003.

[Lan98] Lange, D. B.; Mitsuru, O.: *Programming and Deploying Java Mobile Agents Aglets*, Addison-Wesley Longman Publishing Co., Inc., Boston, MA, USA, 1998.

Bibliography

[Lew02]   Lewis, M.; Jacobson, J.: *Game engines in scientific research - Introduction,* *Communications of the ACM,* Bd. 45, Nr. 1, 2002, S. 27–31.

[Lin99]   Lindholm, T.; Yellin, F.: *The Java Virtual Machine Specification,* Addison-Wesley, Second. Ausg., 1999.

[Lin01]   Lind, J.: *Issues in agent-oriented software engineering,* in *First international workshop, AOSE 2000 on Agent-oriented software engineering,* Springer-Verlag New York, Inc., Secaucus, NJ, USA, 2001, S. 45–58.

[Lös05]   Löser, A.: *Adaptive Overlays in Peer-to-Peer Netzwerken,* PhD thesis, Technische Universität Berlin, Fakultät für Elektrotechnik und Informatik, Sep. 2005.

[Luc03]   Luck, M.; McBurney, P.; Preist, C.: *Agent Technology: Enabling Next Generation Computing (A Roadmap for Agent Based Computing),* AgentLink, 2003.

[Luc04]   Luck, M.: *Guest Editorial: Challenges for Agent-Based Computing.,* *Autonomous Agents and Multi-Agent Systems,* Bd. 9, Nr. 3, 2004, S. 199–201.

[Mag98]   Magnin, L.: *SIEME: An Interactions Based Simulation Model.,* in *ESM,* 1998, S. 410–414.

[Man05]   Manfred, A. D.: *Range Queries in Trie-Structured Overlays,* 2005.

[Mar06]   Marcel Karnstedt and Kai-Uwe Sattler and Manfred Hauswirth and Roman Schmidt, : *Similarity Queries on Structured Data in Structured Overlays,* in *2nd IEEE International Workshop on Networking Meets Databases (NetDB'06),* Atlanta, GA, USA, April 2006.

[mas00]   *Object Management Group – Mobile agent facility specification,* 2000.

[Meh03]   Mehdi, Q. H.; Gough, N. E.; Natkine, S. (Hrsg.): *4th International Conference on Intelligent Games and Simulation (GAME-ON 2003), 19-21 November 2003, London, UK,* EUROSIS, 2003.

[Mer05]   Merugu, S.; Srinivasan, S.; Zegura, E.: *Adding structure to unstructured peer-to-peer networks: the use of small-world graphs,* J. Parallel Distrib. Comput., Bd. 65, Nr. 2, 2005, S. 142–153.

[Mil04]   Milojicic, D. S.; Kalogeraki, V.; Lukose, R.; Nagaraja, K.; Pruyne, J.; Richard, B.; Rollins, S.; Xu, Z.: *Peer-to-Peer Computing,* 2004.

[Mül97]   Müller, J. P.; Wooldridge, M.; Jennings, N. R. (Hrsg.): *Intelligent Agents III, Agent Theories, Architectures, and Languages, ECAI '96 Workshop (ATAL), Budapest, Hungary, August 12-13, 1996, Proceedings,* Bd. 1193 von *Lecture Notes in Computer Science,* Springer, 1997.

[Nap03]   Napster, : *Napster Company Website.,* http://www.nnapster.com, 2003.

[Nej04]   Nejdl, W.; Wolpers, M.; Siberski, W.; Schmitz, C.; Schlosser, M.; Brunkhorst, I.; Loser, A.: *Super-peer-based routing strategies for RDF-based peer-to-peer networks,* 2004.

103

[Pec98]   Pechoucek, M.; Marík, V.; Stepánková, O.; Hazdra, T.: *Tri-Base Model: An Approach to Project-Oriented Production Modelling.*, in *ESM*, 1998, S. 622–626.

[Per97]   Perkins, C. E.; Alpert, S. R.; Woolf, B.: *Mobile IP; Design Principles and Practices*, Addison-Wesley Longman Publishing Co., Inc., Boston, MA, USA, 1997.

[Pet06]   Peterson, L.; Muir, S.; Roscoe, T.; Klingaman, A.: *PlanetLab Architecture: An Overview*, PDN–06–031, PlanetLab Consortium, May 2006.

[Pit01]   Pitt, E.; McNiff, K.: *Java.rmi: The Remote Method Invocation Guide*, Addison-Wesley Longman Publishing Co., Inc., Boston, MA, USA, 2001.

[Piv05]   Pivinen, N.: *Clustering with a minimum spanning tree of scale-free-like structure*, Pattern Recogn. Lett., Bd. 26, Nr. 7, 2005, S. 921–930.

[Rub99]   Rubinstein, M. G.; Duarte, O. C. M. B.: *Evaluating Tradeoffs of Mobile Agents in Network Management*, Networking and Information Systems Journal, Bd. 2, Nr. 2, 1999, S. 237–252, HERMES Science Publications.

[Sei03]   Seigneur, J.-M.; Biegel, G.; Jensen, C. D.: *P2P with JXTA-Java pipes*, in *PPPJ '03: Proceedings of the 2nd international conference on Principles and practice of programming in Java*, Computer Science Press, Inc., New York, NY, USA, 2003, S. 207–212.

[Sie04]   Sierra, C.: *Agent-Mediated Electronic Commerce.*, Autonomous Agents and Multi-Agent Systems, Bd. 9, Nr. 3, 2004, S. 285–301.

[Spr05]   Spranger, S.: *Dynamisches Management von Applikationsdiensten in einer selbstorganisierenden Middleware für MAS*, 2005, Studienarbeit, Friedrich-Schiller Universität Jena, Fakultät für Mathematik und Informatik.

[Sta03]   Staab, S.; Heylighen, F.; Gershenson, C.; Flake, G. W.; Pennock, D. M.; Fain, D. C.; Roure, D. D.; Aberer, K.; Shen, W.-M.; Dousse, O.; Thiran, P.: *Neurons, Viscose Fluids, Freshwater Polyp Hydra-and Self-Organizing Information Systems*, IEEE Intelligent Systems, Bd. 18, Nr. 4, 2003, S. 72–86.

[Ste04]   Stein, E.: *Taschenbuch Rechnernetze und Internet*, Fachbuchverlag Leipzig, 2., bearb. aufl.. Ausg., 2004.

[Sto96]   Stonebraker, M.; Aoki, P. M.; Litwin, W.; Pfeffer, A.; Sah, A.; Sidell, J.; Staelin, C.; Yu, A.: , Bd. 5, Nr. 1, Jan. 1996, S. 48–63, Electronic edition.

[try06]   *Tryllian Solutions – Agent Development Kit (ADK) 3.2.0*, http://www.tryllian.com/technology/product1.html, 2006, [Online; accessed 14-June-2006].

[Var05]   Varela, C. A.; Ciancarini, P.; Taura, K.: *Worldwide computing: Adaptive middleware and programming technology for dynamic Grid environments*, Scientific Programming Journal, Bd. 13, Nr. 4, December 2005, S. 255–263, Guest Editorial.

Bibliography

[Vou01]   Voulgaris, S.; Kermarrec, A.; Massoulié, L.; van Steen, M.: *Exploiting Semantic Proximity in Peer-to-Peer Content Searching*, in *10th International Workshop on Future Trends in Distributed Computing Systems (FTDCS 2004)*, Nov. 2001.

[Wik06]   Wikipedia, : *Software agent — Wikipedia, The Free Encyclopedia*, http://en.wikipedia.org/w/index.php?title=Software_agent&oldid=55958249, 2006, [Online; accessed 7-June-2006].

[ZY04]   Zeinalipour-Yazti, D.; Kalogeraki, V.; Gunopulos, D.: *Information Retrieval Techniques for Peer-to-Peer Networks*, *Computing in Science and Engineering*, Bd. 06, Nr. 4, 2004, S. 20–26.

List of Figures

www.ingramcontent.com/pod-product-compliance
Lightning Source LLC
Chambersburg PA
CBHW052016230326
41598CB00078B/3490